W9-AEU-932

STATE VARIABLE METHODS IN AUTOMATIC CONTROL

STATE VARIABLE METHODS IN AUTOMATIC CONTROL

KATSUHISA FURUTA
Tokyo Institute of Technology

AKIRA SANO
Keio University, Yokohama

In association with

DEREK ATHERTON
Sussex University, UK

JOHN WILEY & SONS
Chichester · New York · Brisbane · Toronto · Singapore

This work is an extended version of that first
published by Corona Publishing Company Limited,
4–46 10-Sengoku Bunkyo-ku Tokyo, Japan

Library of Congress Cataloging-in-Publication Data:

Furuta, Katsuhisa, 1940–
State variable methods in automatic control / Katsuhisa Furuta,
Akira Sano ; in association with Derek Atherton.
p. cm.
ISBN 0 471 91877 6
1. System analysis. 2. Control theory. 3. Mathematical
optimization. I. Sano, Akira, 1943– . II. Atherton, Derek P.
III. Title.
QA402.F87 1988
629.8′3—dc 19 87-35481
CIP

British Library Cataloguing in Publication Data:

Furuta, Katsuhisa
State variable methods in automatic control.
1. Automatic control
I. Title II. Sano, Akira III. Atherton,
Derek
629.8′32 TJ213

ISBN 0 471 91877 6

Typeset by Mathematical Composition Setters Ltd, Ivy Street, Salisbury
Printed and bound in Great Britain by Biddles of Guildford

CONTENTS

1

MATHEMATICAL DESCRIPTION OF LINEAR SYSTEMS

1.1 SYSTEM REPRESENTATION

1.1.1 Input–Output Description

The input–output description of a system gives a mathematical relationship between the input and output of the system. The impulse response, the transfer function and the frequency response are typical examples of this description. The system shown in Figure 1.1 has m inputs and p outputs, which are written as the vectors $\mathbf{u}(t) = (u_1, \ldots, u_m)^{\mathrm{T}}$ and $\mathbf{y}(t) = (y_1, \ldots, y_p)^{\mathrm{T}}$ respectively. If the time function $\mathbf{u}$ is defined only over the interval $[t_0, t_1]$, we write it as $\mathbf{u}_{[t_0, t_1]}$. The output at time t of the system generally depends not only on the input applied at t, but also on the input before and/or after t. Therefore the input–output description is given by

$$\begin{aligned}\mathbf{y}(t) &= \mathbf{y}(t, \mathbf{u}_{(-\infty,\infty)}) \\ &= \int_{-\infty}^{\infty} H(t, \tau)\mathbf{u}(\tau)\, \mathrm{d}\tau \qquad (1.1)\end{aligned}$$

where $H(t, \tau)$ is the matrix impulse response, and the (i, j) element $h_{ij}(t, \tau)$ is the impulse response associated with the input $u_j(t)$ and the output $y_i(t)$. The physical interpretation of $h_{ij}(t, \tau)$ is observed from the fact that when the impulse function $u_j(t) = \delta(t - \tau)$ is applied as an input at time $t = \tau$ the output is represented by $y_j(t) = h_{ij}(t, \tau)$.

If the characteristics of a system do not change with time, the system is said to be *time-invariant* or *stationary*. The time-invariant linear system has the property that if an input is shifted by an amount of time, the waveform of the output remains the same except for the shift by the same amount of time. Therefore, the impulse response $H(t, \tau)$ depends only on the difference $t - \tau$, and the input–output description is given by

$$\mathbf{y}(t) = \int_{-\infty}^{\infty} H(t - \tau)\mathbf{u}(\tau)\, \mathrm{d}\tau \qquad (1.2)$$

Generally in dynamical systems, the output at time t depends not only on the input at t but also on the input applied before and/or after t. The system

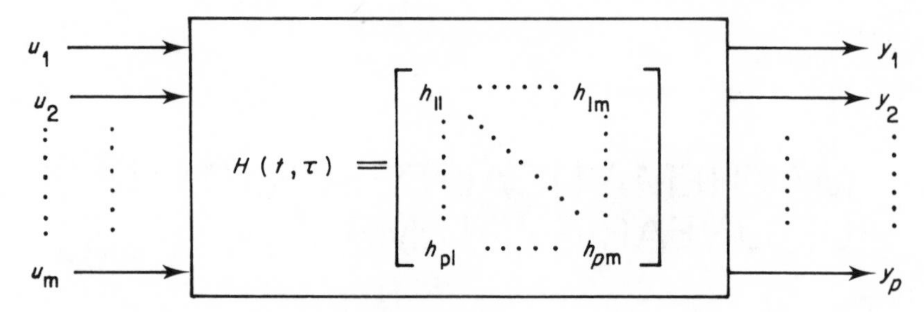

Figure 1.1 Matrix impulse response function

is said to be *causal* if the output at t does not depend on the input after t, that is, it depends only on the input applied before and at time t. Every physically realizable system satisfies the causality property. The impulse response of a causal system therefore satisfies

$$H(t, \tau) = 0 \qquad \text{for } t < \tau \tag{1.3}$$

and hence the input–output description is given by

$$\begin{aligned}\mathbf{y}(t) &= \mathbf{y}(t, \mathbf{u}_{(-\infty, t]}) \\ &= \int_{-\infty}^{t} H(t, \tau)\mathbf{u}(\tau)\,\mathrm{d}\tau \end{aligned} \tag{1.4}$$

1.1.2 State Variable Description

We consider the output response $\mathbf{y}(t)$ after time t_0 of a system excited by the input $\mathbf{u}_{[t_0,\infty)}$. Generally the output $\mathbf{y}(t)$ for $t \geqslant t_0$ cannot be uniquely determined by knowledge of the input $\mathbf{u}_{[t_0,\infty)}$ alone. Rewriting (1.1) yields

$$\mathbf{y}(t) = \int_{-\infty}^{t_0} H(t, \tau)\mathbf{u}(\tau)\,\mathrm{d}\tau + \int_{t_0}^{\infty} H(t, \tau)\mathbf{u}(\tau)\,\mathrm{d}\tau \tag{1.5}$$

and it is noticed that the output $y(t)$ is excited not only by the input $\mathbf{u}_{[t_0,\infty)}$ but also by the unknown input $\mathbf{u}_{(-\infty,t_0)}$ before t_0. The first term in the RHS of (1.5) corresponds to the initial condition at time t_0. If the n-tuple parameters $\mathbf{x} = (x_1, \ldots, x_n)^{\mathrm{T}}$ which specify the initial condition are given together with the future input $\mathbf{u}_{[t_0,\infty)}$, the output $\mathbf{y}(t)$ can be uniquely calculated. $\mathbf{x}$ is a function of t_0, and we can represent the output $\mathbf{y}(t)$ by

$$\mathbf{y}(t) = \mathbf{y}(t, \mathbf{x}(t_0), \mathbf{u}_{[t_0,\infty)}) \qquad \text{for } t \geqslant t_0 \tag{1.6}$$

where $\mathbf{x}(t_0)$ is called the *state* or *state variable* at time t_0, and n is the order of the system. The state is the minimal sufficient information that enables us to uniquely calculate the future output $\mathbf{y}(t)$ on $t \geqslant t_0$ for the input $\mathbf{u}_{[t_0,\infty)}$ without reference to the past input $\mathbf{u}_{(-\infty,t_0)}$ before time t_0.

Example 1.1 Consider the series RLC circuit shown in Figure 1.2, where the input is the voltage $u(t)$ and the output is the capacitor voltage $y(t)$. The transfer function from u to y is easily evaluated to be

$$H(s) = \frac{Y(s)}{U(s)} = \frac{1/s}{(3 + 2s + 1/s)} = \frac{1}{s + 0.5} - \frac{1}{s + 1}$$

and the impulse response $h(t)$ is given by

$$h(t) = \mathscr{L}^{-1}H(s) = \mathrm{e}^{-0.5t} - \mathrm{e}^{-t}.$$

Denoting the initial time as t_0 and taking account of the input $u_{[t_0,t]}$, the output $y(t)$ is described as follows:

$$\begin{aligned} y(t) &= \int_{-\infty}^{t} h(t-\tau)u(\tau)\,\mathrm{d}\tau \\ &= \int_{-\infty}^{t_0} h(t-\tau)u(\tau)\,\mathrm{d}\tau + \int_{t_0}^{t} h(t-\tau)u(\tau)\,\mathrm{d}\tau. \end{aligned} \tag{1.7}$$

The first term in the RHS of (1.7) indicates the effect of the unknown input before t_0 on the output at t, and is rewritten as

$$\begin{aligned} \int_{-\infty}^{t_0} h(t-\tau)u(\tau)\,\mathrm{d}\tau &= \mathrm{e}^{-0.5t}\int_{-\infty}^{t_0} \mathrm{e}^{0.5\tau}u(\tau)\,\mathrm{d}\tau - \mathrm{e}^{-t}\int_{-\infty}^{t_0} \mathrm{e}^{\tau}u(\tau)\,\mathrm{d}\tau \\ &= \mathrm{e}^{-0.5t}x_1 - \mathrm{e}^{-t}x_2 \end{aligned} \tag{1.8}$$

where

$$x_1 = x_1(t_0) = \int_{-\infty}^{t_0} \mathrm{e}^{0.5\tau}u(\tau)\,\mathrm{d}\tau$$

$$x_2 = x_2(t_0) = \int_{-\infty}^{t_0} e^{\tau}u(\tau)\,\mathrm{d}\tau$$

x_1 and x_2 are both independent of t. Thus if x_1 and x_2 are known, the output $y(t)$ for $t \geqslant t_0$ when the circuit is excited by the input $u_{[t_0,t]}$ can be uniquely calculated. Hence x_1 and x_2 are the minimum amount of information at t_0 for determining the future output response and therefore we can assign $x_1(t)$ and $x_2(t)$ as the states for any time t.

It follows from (1.7) and (1.8) that

$$y(t_0) = \mathrm{e}^{-0.5t_0}x_1 - \mathrm{e}^{-t_0}x_2. \tag{1.9}$$

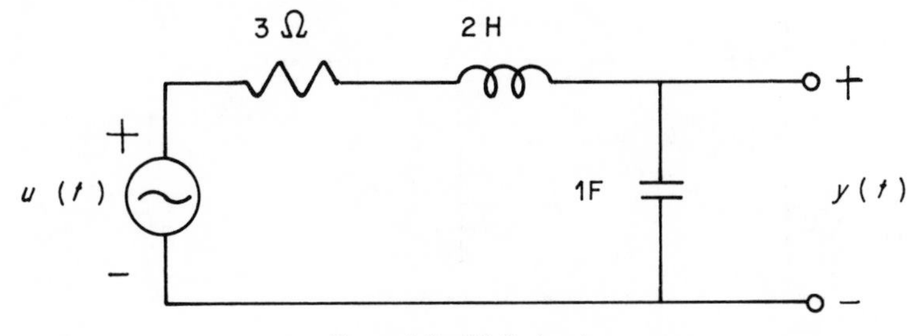

Figure 1.2 RLC circuit

Differentiating (1.7) with respect to t yields

$$\dot{y}(t) = -0.5\,\mathrm{e}^{-0.5t}x_1 + \mathrm{e}^{-t}x_2 + h(0)u(t) + \int_{t_0}^{t} \frac{\partial}{\partial t} h(t-\tau)u(\tau)\,\mathrm{d}\tau,$$

and letting $t = t_0$ in the above equation gives

$$\dot{y}(t_0) = -0.5\,\mathrm{e}^{-0.5t_0}x_1 + \mathrm{e}^{-t_0}x_2. \tag{1.10}$$

From (1.9) and (1.10) we can write

$$x_1 = 2\,\mathrm{e}^{0.5t_0}[y(t_0) + \dot{y}(t_0)]$$
$$x_2 = \mathrm{e}^{t_0}[y(t_0) + 2\dot{y}(t_0)].$$

This implies that $y(t_0)$ and $\dot{y}(t_0)$ can also be chosen as state variables. They are the physical quantities related with the capacitor charge and the inductor current respectively. It is thus important to note that the choice of the state variables is not unique. We will later show a standard method for deriving the state equation for a general RLC network, where capacitor voltages and inductor currents are taken as the state variables.

Putting $t = t_0$ in (1.4), we have

$$\begin{aligned}\mathbf{y}(t) &= \mathbf{y}[t, \mathbf{x}(t), \mathbf{u}_{[t,t]})] \\ &= \mathbf{g}[t, \mathbf{x}(t), \mathbf{u}(t)]\end{aligned} \tag{1.11}$$

This implies that the output $\mathbf{y}(t)$ can be specified by the state $\mathbf{x}(t)$ and the input $\mathbf{u}(t)$ at time t. We call $\mathbf{g}(\cdot)$ the *output function*.

On the other hand, from (1.6) and (1.11) it is easy to see that $\mathbf{x}(t)$ depends on $\mathbf{x}(t_0)$ and $\mathbf{u}_{[t_0,t]}$ and is hence represented as

$$\mathbf{x}(t) = \psi(t;\, \mathbf{x}_0, t_0, \mathbf{u}_{[t_0,t]}) \qquad \text{for } \mathbf{x}_0 = \mathbf{x}(t_0) \tag{1.12}$$

where $\psi(\cdot)$ is called the *state transition function*, since it indicates the transition behaviour from the initial state $\mathbf{x}_0$ to the state $\mathbf{x}(t)$ by the application of the input $\mathbf{u}_{[t_0,t]}$.

In summary, a dynamical system can be described in the compact form shown in Figure 1.3, by using the state-transition function (1.12), the output function (1.11) and incorporating the concept of state variables.

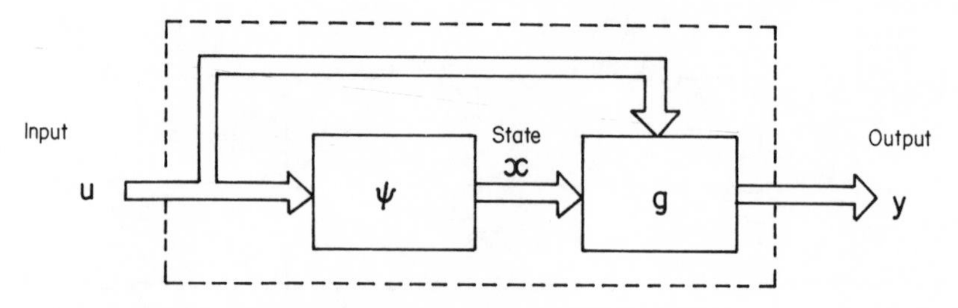

Figure 1.3 Schematic diagram of a dynamical system

We treat, in many cases, the transition function $\psi(\cdot)$ given by the solution of the ordinary differential equation:

$$\dot{\mathbf{x}}(t) = \mathbf{f}[\mathbf{x}(t), \mathbf{u}(t), t] \qquad \text{for } \mathbf{x}(t_0) = \mathbf{x}_0 \tag{1.13}$$

Combined with the output equation (1.11), the description of the dynamical system is summarized as

$$\begin{aligned} \dot{\mathbf{x}}(t) &= \mathbf{f}[\mathbf{x}(t), \mathbf{u}(t), t] \qquad \mathbf{x}(t_0) = \mathbf{x}_0 \\ \mathbf{y}(t) &= \mathbf{g}[\mathbf{x}(t), \mathbf{u}(t), t] \end{aligned} \tag{1.14}$$

Provided that $\mathbf{f}(\cdot)$ satisfies the Lipschitz condition with respect to $\mathbf{x}$ and is continuous for $\mathbf{u}$ and t, and that $\mathbf{g}(\cdot)$ is continuous for $\mathbf{x}$, $\mathbf{u}$ and t, the state $\mathbf{x}(t)$ and the output $\mathbf{y}(t)$ uniquely exist from any initial condition $\mathbf{x}_0$ and input $\mathbf{u}(t)$. $\mathbf{x}$, $\mathbf{u}$ and $\mathbf{y}$ are vectors with n, m and p elements respectively, and the space spanned by $\mathbf{x}$ is called the n-dimensional *state space*. If $\mathbf{f}(\cdot)$ and $\mathbf{g}(\cdot)$ in (1.14) do not include t explicitly then the dynamical system is *time-invariant* or *stationary*.

If $\mathbf{f}(\cdot)$ and $\mathbf{g}(\cdot)$ are linear with respect to $\mathbf{x}$ and $\mathbf{u}$, the linear dynamical system is described by:

(a) Time-variant linear system $[A(t), B(t), C(t), D(t)]$:

$$\dot{\mathbf{x}}(t) = A(t)\mathbf{x}(t) + B(t)\mathbf{u}(t) \qquad \text{for } \mathbf{x}(t_0) = \mathbf{x}_0 \tag{1.15a}$$

$$\mathbf{y}(t) = C(t)\mathbf{x}(t) + D(t)\mathbf{u}(t) \tag{1.15b}$$

(b) Time-invariant linear system (A, B, C, D):

$$\dot{\mathbf{x}}(t) = A\mathbf{x}(t) + B\mathbf{u}(t) \qquad \text{for } \mathbf{x}(t_0) = \mathbf{x}_0 \tag{1.16a}$$

$$\mathbf{y}(t) = C\mathbf{x}(t) + D\mathbf{u}(t) \tag{1.16b}$$

where $A(t)$, $B(t)$, $C(t)$, and $D(t)$ (or A, B, C, and D) are matrices with $n \times n$, $n \times m$, $p \times n$ and $p \times m$ elements respectively. If the elements are continuous with respect to t then there exists a unique solution of (1.15).

1.1.3 Relationship between Transfer Function and State Variable Description

Another description of the input–output relationship is the Laplace transform of the impulse response matrix which we call the *transfer function matrix*. If the first term in the RHS of (1.5) is assumed zero, the output $\mathbf{y}(t)$ depends only on the input $\mathbf{u}_{[t_0,t]}$. In the time-invariant causal linear system the input–output description, assuming $t_0 = 0$, is

$$\mathbf{y}(t) = \int_0^\infty H(t-\tau)\mathbf{u}(\tau)\,\mathrm{d}\tau \tag{1.17}$$

Let the Laplace transform of $H(t), \mathbf{y}(t)$, and $\mathbf{u}(t)$ be defined by

$$H(s) = \mathscr{L}\{H(t)\} = \int_0^\infty e^{-st} H(t)\, dt$$
$$\mathbf{Y}(s) = \mathscr{L}\{\mathbf{y}(t)\}, \qquad \mathbf{U}(s) = \mathscr{L}\{\mathbf{u}(t)\} \tag{1.18}$$

The application of the Laplace transformation to (1.17) gives

$$\mathbf{Y}(s) = H(s)\mathbf{U}(s) \tag{1.19}$$

where $H(s)$ is called the matrix transfer function.

We now determine the transfer function of the linear system (A, B, C, D) described by (1.16). Applying the Laplace transform to (1.16) leads to

$$s\mathbf{X}(s) - \mathbf{x}_0 = A\mathbf{X}(s) + B\mathbf{U}(s) \tag{1.20a}$$
$$\mathbf{Y}(s) = C\mathbf{X}(s) + D\mathbf{U}(s) \tag{1.20b}$$

where $\mathbf{X}(s) = \mathscr{L}\{\mathbf{x}(t)\}, \mathbf{U}(s) = \mathscr{L}\{\mathbf{u}(t)\}$ and $\mathbf{Y}(s) = \mathscr{L}\{\mathbf{y}(t)\}$. Rewriting (1.20) we have

$$\mathbf{X}(s) = (sI - A)^{-1}\mathbf{x}_0 + (sI - A)^{-1}B\mathbf{U}(s) \tag{1.21a}$$
$$\mathbf{Y}(s) = C(sI - A)^{-1}\mathbf{x}_0 + \{C(sI - A)^{-1}B + D\}\mathbf{U}(s) \tag{1.21b}$$

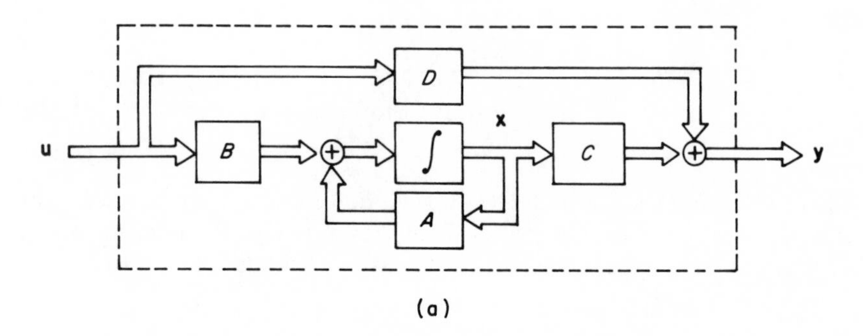

(a)

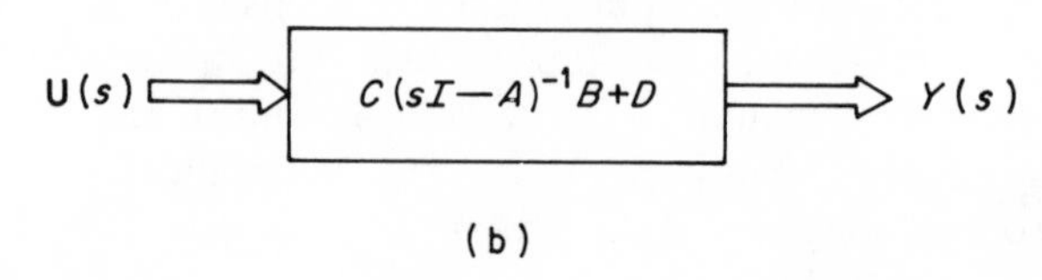

(b)

Figure 1.4 Block diagrams of a linear dynamical system. (a) state variable description, and (b) input–output description (transfer matrix function)

where I is the identity matrix. In the case when the initial condition is zero, i.e. $\mathbf{x}_0 = 0$, the transfer function from $\mathbf{u}$ to $\mathbf{y}$ is seen to be given by

$$H(s) = C(sI - A)^{-1}B + D \tag{1.22}$$

$$= \frac{C\,\mathrm{adj}(sI - A)B}{\det(sI - A)} + D \tag{1.23}$$

The relation between the descriptions $H(s)$ and (A, B, C, D) is illustrated in Figure 1.4.

Example 1.2 Consider the linear single-input single-output (SISO) system $(A, \mathbf{b}, \mathbf{c}^{\mathrm{T}}, 0)$† described by

$$\begin{pmatrix} \dot{x}_1 \\ \dot{x}_2 \\ \dot{x}_3 \end{pmatrix} = \begin{pmatrix} 0 & 1 & 0 \\ 0 & -1 & -1 \\ 0 & 0 & -3 \end{pmatrix} \begin{pmatrix} x_1 \\ x_2 \\ x_3 \end{pmatrix} + \begin{pmatrix} 0 \\ 1 \\ 1 \end{pmatrix} u$$

$$y = (1 \quad 0 \quad 0)\mathbf{x}$$

Then,

$$\det(sI - A) = \det \begin{pmatrix} s & -1 & 0 \\ 0 & s+1 & 1 \\ 0 & 0 & s+3 \end{pmatrix} = s(s+1)(s+3)$$

The transfer function $H(s)$ is given from (1.23) by

$$H(s) = \frac{Y(s)}{U(s)} = \frac{(1 \quad 0 \quad 0)\begin{pmatrix} (s+1)(s+3) & s+3 & -1 \\ 0 & s(s+3) & -s \\ 0 & 0 & s(s+1) \end{pmatrix}\begin{pmatrix} 0 \\ 1 \\ 1 \end{pmatrix}}{s(s+1)(s+3)}$$

$$= \frac{s+2}{s(s+1)(s+3)}$$

There are many methods to compute $(sI - A)^{-1}$. One is to make use of (1.23), which requires calculations of determinants in both the numerator and denominator as done above. Here we present an iterative scheme to calculate $(sI - A)^{-1}$, which is called the *Faddeev algorithm*.

Let the characteristic polynomial of the $n \times n$ matrix A be denoted by

$$\det(sI - A) \equiv \phi(s)$$

$$= s^n + \alpha_{n-1}s^{n-1} + \alpha_{n-2}s^{n-2} + \cdots + \alpha_0 \tag{1.24}$$

Then,

$$(sI - A)^{-1} = \frac{1}{\phi(s)}(\Gamma_{n-1}s^{n-1} + \Gamma_{n-2}s^{n-2} + \cdots + \Gamma_0) \tag{1.25}$$

†This representation is used since for a SISO system B becomes a column vector and C a row vector

where $\alpha_{n-1}, \alpha_{n-2}, \ldots, \alpha_0, \Gamma_{n-1}, \Gamma_{n-2}, \ldots, \Gamma_0$ can be calculated in a recursive manner from:

$$
\begin{aligned}
\Gamma_{n-1} &= I & \alpha_{n-1} &= -\operatorname{tr}(A\Gamma_{n-1}) \\
\Gamma_{n-2} &= A\Gamma_{n-1} + \alpha_{n-1}I & \alpha_{n-2} &= -\operatorname{tr}(A\Gamma_{n-2})/2 \\
\Gamma_{n-3} &= A\Gamma_{n-2} + \alpha_{n-2}I & \alpha_{n-3} &= -\operatorname{tr}(A\Gamma_{n-3})/3 \\
&\cdots & &\cdots \\
\Gamma_i &= A\Gamma_{i+1} + \alpha_{i+1}I & \alpha_i &= -\operatorname{tr}(A\Gamma_i)/(n-i) \\
&\cdots & &\cdots \\
\Gamma_0 &= A\Gamma_1 + \alpha_1 I & \alpha_0 &= -\operatorname{tr}(A\Gamma_0)/n
\end{aligned}
\tag{1.26a}
$$

$$0 = A\Gamma_0 + \alpha_0 I \tag{1.26b}$$

where tr(X), the trace of X, is the sum of all the diagonal elements of the matrix X.

Example 1.3 Here we again compute $(sI - A)^{-1}$ which appeared in Example 1.2 but this time using the Faddeev algorithm (1.26).

$$\Gamma_2 = I, \qquad \alpha_2 = -\operatorname{tr}(A) = 4$$

$$\Gamma_1 = A\Gamma_2 + 4I = \begin{pmatrix} 4 & 1 & 0 \\ 0 & 3 & -1 \\ 0 & 0 & 1 \end{pmatrix}, \qquad \alpha_1 = -\operatorname{tr}(A\Gamma_1)/2 = 3$$

$$\Gamma_0 = A\Gamma_1 + 3I = \begin{pmatrix} 3 & 3 & -1 \\ 0 & 0 & 0 \\ 0 & 0 & 0 \end{pmatrix}, \qquad \alpha_0 = -\operatorname{tr}(A\Gamma_0)/3 = 0$$

Hence from (1.24) and (1.25), we obtain

$$(sI - A)^{-1} = \frac{1}{s^3 + 4s^2 + 3s} \begin{pmatrix} s^2 + 4s + 3 & s + 3 & -1 \\ 0 & s^2 + 3s & -s \\ 0 & 0 & s^2 + s \end{pmatrix}$$

The recursive form for $\{\Gamma_i\}$ in (1.26) can be derived as follows: Rearranging (1.25), we obtain

$$\phi(s)I = (sI - A)(\Gamma_{n-1}s^{n-1} + \Gamma_{n-2}s^{n-2} + \cdots + \Gamma_1 s + \Gamma_0)$$

and equating the coefficients with the same power of s yields the formula for $\{\Gamma_i\}$ in (1.26).

Eliminating $\Gamma_{n-1}, \Gamma_{n-2}, \ldots, \Gamma_0$ from (1.26) iteratively, we have

$$A^n + \alpha_{n-1}A^{n-1} + \alpha_{n-2}A^{n-1} + \cdots + \alpha_1 A + \alpha_0 I = 0 \tag{1.27}$$

which also implies $\phi(A) = 0$. This well known result is the *Cayley-Hamilton theorem*, which will be frequently referred to later. It is easily shown from this theorem that, for any integer $i > 0$, A^{n+i} can be expressed as a linear combination of $I, A, \ldots, A^{n-1}$.

1.2 STATE VARIABLE MODELLING

1.2.1 Linearization

The technique of linearizing nonlinear systems is extremely important since it provides in many cases a standard approach to analysis. For instance, the method is commonly employed in process control where a process is regulated around certain set points. The linearization technique is also used in designing a control system so as to keep the output along the nominal trajectory.

It is now assumed that a nonlinear control system is described by the differential equation:

$$\dot{\mathbf{x}}(t) = \mathbf{f}(\mathbf{x}(t), \mathbf{u}(t)), \qquad \mathbf{x}(t_0) = \mathbf{x}_0 \tag{1.28a}$$

$$\mathbf{y}(t) = \mathbf{g}[\mathbf{x}(t), \mathbf{u}(t)], \tag{1.28b}$$

If the nominal state vector is defined by $\mathbf{x}^*(t)$ and the corresponding control input by $\mathbf{u}^*(t)$, the state $\mathbf{x}^*(t)$ satisfies

$$\begin{aligned} \dot{\mathbf{x}}^*(t) &= \mathbf{f}[\mathbf{x}^*(t), \mathbf{u}^*(t)] \\ \mathbf{y}^*(t) &= \mathbf{g}[\mathbf{x}^*(t), \mathbf{u}^*(t)] \end{aligned} \tag{1.29}$$

We now consider the deviation of the state and the output from their nominal trajectory owing to the fact that the input deviates from $\mathbf{u}^*(t)$. We define these deviations by

$$\begin{aligned} \delta\mathbf{u}(t) &= \mathbf{u}(t) - \mathbf{u}^*(t) \\ \delta\mathbf{x}(t) &= \mathbf{x}(t) - \mathbf{x}^*(t) \\ \delta\mathbf{y}(t) &= \mathbf{y}(t) - \mathbf{y}^*(t) \end{aligned}$$

If these variations are assumed to be small, we can expand (1.28) in a Taylor series around the nominal value, as follows

$$\begin{aligned} \dot{x}_i^*(t) + \delta\dot{x}_i^*(t) = f_i[\mathbf{x}^*(t), \mathbf{u}^*(t)] &+ \left.\frac{\partial f_i(\mathbf{x}, \mathbf{u})}{\partial \mathbf{x}}\right|^{\mathrm{T}}_{\substack{\mathbf{x}=\mathbf{x}^* \\ \mathbf{u}=\mathbf{u}^*}} \delta\mathbf{x}(t) \\ &+ \left.\frac{\partial f_i(\mathbf{x}, \mathbf{u})}{\partial \mathbf{u}}\right|^{\mathrm{T}}_{\substack{\mathbf{x}=\mathbf{x}^* \\ \mathbf{u}=\mathbf{u}^*}} \delta\mathbf{u}(t) + o(\delta\mathbf{x}, \delta\mathbf{u}) \end{aligned}$$

for $i = 1, 2, \ldots, n$

$$y_i^*(t) + \delta y_i(t) = g_i[\mathbf{x}^*(t), \mathbf{u}^*(t)] + \left.\frac{\partial g_i(\mathbf{x}, \mathbf{u})}{\partial \mathbf{x}}\right|^{\mathrm{T}}_{\substack{\mathbf{x}=\mathbf{x}^* \\ \mathbf{u}=\mathbf{u}^*}} \delta\mathbf{x}(t)$$

$$+ \left.\frac{\partial g_i(\mathbf{x}, \mathbf{u})}{\partial \mathbf{u}}\right|^{\mathrm{T}}_{\substack{\mathbf{x}=\mathbf{x}^* \\ \mathbf{u}=\mathbf{u}^*}} \delta\mathbf{u}(t) + o(\delta\mathbf{x}, \delta\mathbf{u})$$

$$\text{for } i = 1, 2, \ldots, p$$

where $o(\delta\mathbf{x}, \delta\mathbf{u})$ denotes higher order terms.

By introducing the Jacobian matrices

$$A(t) \equiv \left.\frac{\partial \mathbf{f}(\mathbf{x}, \mathbf{u})}{\partial \mathbf{x}^{\mathrm{T}}}\right|_{\substack{\mathbf{x}=\mathbf{x}^* \\ \mathbf{u}=\mathbf{u}^*}} = \left.\begin{pmatrix} \dfrac{\partial f_1}{\partial x_1} & \cdots & \dfrac{\partial f_1}{\partial x_n} \\ \vdots & & \vdots \\ \dfrac{\partial f_n}{\partial x_1} & \cdots & \dfrac{\partial f_n}{\partial x_n} \end{pmatrix}\right|_{\substack{\mathbf{x}=\mathbf{x}^* \\ \mathbf{u}=\mathbf{u}^*}}$$

$$B(t) \equiv \left.\frac{\partial \mathbf{f}(\mathbf{x}, \mathbf{u})}{\partial \mathbf{u}^{\mathrm{T}}}\right|_{\substack{\mathbf{x}=\mathbf{x}^* \\ \mathbf{u}=\mathbf{u}^*}} = \left.\begin{pmatrix} \dfrac{\partial f_1}{\partial u_1} & \cdots & \dfrac{\partial f_1}{\partial u_m} \\ \vdots & & \vdots \\ \dfrac{\partial f_n}{\partial u_1} & \cdots & \dfrac{\partial f_n}{\partial u_m} \end{pmatrix}\right|_{\substack{\mathbf{x}=\mathbf{x}^* \\ \mathbf{u}=\mathbf{u}^*}}$$

$$C(t) \equiv \left.\frac{\partial \mathbf{g}(\mathbf{x}, \mathbf{u})}{\partial \mathbf{x}^{\mathrm{T}}}\right|_{\substack{\mathbf{x}=\mathbf{x}^* \\ \mathbf{u}=\mathbf{u}^*}} = \left.\begin{pmatrix} \dfrac{\partial g_1}{\partial x_1} & \cdots & \dfrac{\partial g_1}{\partial x_n} \\ \vdots & & \vdots \\ \dfrac{\partial g_p}{\partial x_1} & \cdots & \dfrac{\partial g_p}{\partial x_n} \end{pmatrix}\right|_{\substack{\mathbf{x}=\mathbf{x}^* \\ \mathbf{u}=\mathbf{u}^*}}$$

$$D(t) \equiv \left.\frac{\partial \mathbf{g}(\mathbf{x}, \mathbf{u})}{\partial \mathbf{u}^{\mathrm{T}}}\right|_{\substack{\mathbf{x}=\mathbf{x}^* \\ \mathbf{u}=\mathbf{u}^*}} = \left.\begin{pmatrix} \dfrac{\partial g_1}{\partial u_1} & \cdots & \dfrac{\partial g_1}{\partial u_m} \\ \vdots & & \vdots \\ \dfrac{\partial g_p}{\partial u_1} & \cdots & \dfrac{\partial g_p}{\partial u_m} \end{pmatrix}\right|_{\substack{\mathbf{x}=\mathbf{x}^* \\ \mathbf{u}=\mathbf{u}^*}}$$

we have the linearized system description

$$\delta\dot{\mathbf{x}}(t) = A(t)\,\delta\mathbf{x}(t) + B(t)\,\delta\mathbf{u}(t), \qquad \text{for } \delta\mathbf{x}(t_0) = \delta\mathbf{x}_0 \tag{1.30a}$$

$$\delta\mathbf{y}(t) = C(t)\,\delta\mathbf{x}(t) + D(t)\,\delta\mathbf{u}(t) \tag{1.30b}$$

This equation describes the variations around the nominal trajectory and has at least first-order accuracy.

If $\mathbf{x}^*$, $\mathbf{u}^*$ and $\mathbf{y}^*$ are chosen constant as the equilibrium or steady state such that they satisfy

$$0 = \mathbf{f}(\mathbf{x}^*, \mathbf{u}^*) \tag{1.31a}$$

$$\mathbf{y}^* = \mathbf{g}(\mathbf{x}^*, \mathbf{u}^*), \tag{1.31b}$$

the linearized equation corresponding to (1.30) becomes the time-invariant linear system

$$\delta\dot{\mathbf{x}}(t) = A\ \delta\mathbf{x}(t) + B\ \delta\mathbf{u}(t) \tag{1.32a}$$

$$\delta\mathbf{y}(t) = C\ \delta\mathbf{x}(t) + D\ \delta\mathbf{u}(t) \tag{1.32b}$$

Example 1.4 (Dynamics of a stirred tank). Consider the stirred tank shown in Figure 1.5. The tank is fed with an incoming flow with constant temperature θ_i (°C) and flow rate is $q_i(t)$ (m^3/s). We consider the dynamical behaviour of the temperature and the level of the tank, which are later chosen as the state variables. It is assumed that the tank is heated and stirred well so that the temperature of the outgoing flow equals the temperature $\theta_o(t)$ in the tank. Other symbols are defined as follows: $q_o(t)$(m^3/s) is the outgoing flow rate, $\nu(t)$ (cal/s) the heat input, c (cal/m^3 °C) the specific heat of the flow, $h(t)$ the level of the tank, S (m^2) the cross-sectional area of the tank and hence the volume, $V(t)$, is equal to $Sh(t)$.

From the mass balance equation, we have

$$S\dot{h}(t) = q_i(t) - q_o(t) \tag{1.33a}$$

$$q_o(t) = \alpha_\searrow[h(t)] \tag{1.33b}$$

$$cS\frac{\mathrm{d}}{\mathrm{d}t}\{\theta_o(t)h(t)\} = c\theta_i q_i(t) - c\theta_o(t)q_o(t) + \nu(t). \tag{1.33c}$$

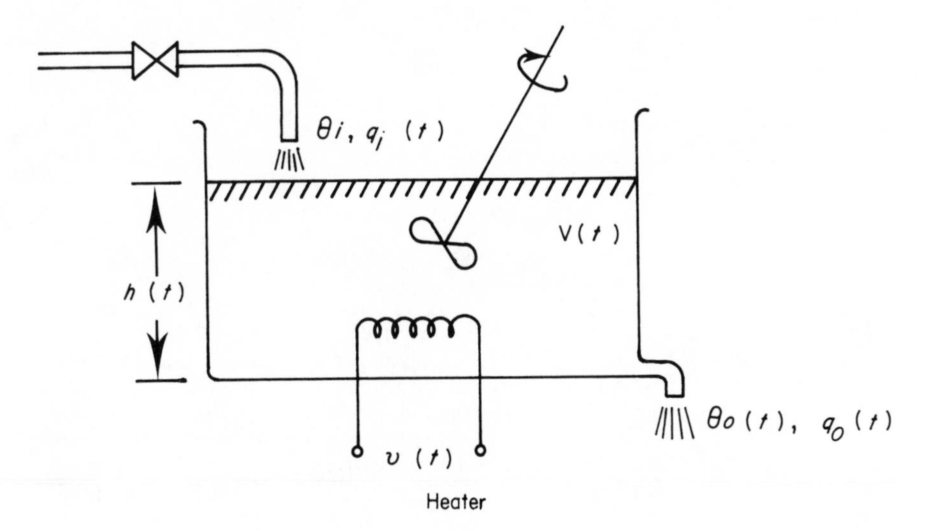

Figure 1.5 Stirred tank

We take $q_i(t)$ and $\nu(t)$ as the inputs and the output $\mathbf{y}(t)$ to be

$$\mathbf{y}(t) = \begin{pmatrix} q_o(t) \\ \theta_o(t) \end{pmatrix} = \begin{pmatrix} \alpha\sqrt{[h(t)]} \\ \theta_o(t) \end{pmatrix} \tag{1.34}$$

We consider now a steady-state situation where all quantities are constant, i.e. q_i^* and q_o^* for the flow rates, ν^* for the heating rate, h^* for the level and θ^* for the temperatures. In this stationary state, the following relations hold:

$$0 = q_i^* - q_o^*, \quad q_o^* = \alpha\sqrt{(h^*)} \tag{1.35a}$$

$$0 = c\theta_i{}^* q_i{}^* - c\theta_o{}^* q_o{}^* + \nu^* \tag{1.35b}$$

and the output $\begin{pmatrix} q_o^* \\ \theta_o^* \end{pmatrix} = \begin{pmatrix} \alpha\sqrt{(h^*)} \\ \theta_o^* \end{pmatrix}$. (1.36)

We now assume only small deviations from the steady state such that

$$q_i(t) = q_i^* + \delta q_i(t), \quad \nu(t) = \nu^* + \delta\nu(t)$$
$$h(t) = h^* + \delta h(t), \quad \theta_o(t) = \theta_o^* + \delta\theta_o(t)$$

and expand (1.33a) to (1.33c) in Taylor series around the stationary values to obtain

$$\delta\dot{h}(t) = -\frac{\alpha\sqrt{(h^*)}}{2Sh^*}\,\delta h(t) + \frac{1}{S}\,\delta q_i(t)$$

$$h^*\,\delta\dot{\theta}_o(t) + \theta_o^*\,\delta\dot{h}(t) = -\frac{\alpha\sqrt{(h^*)}}{2Sh^*}\,\theta_o^*\,\delta h(t) - \frac{\alpha\sqrt{(h^*)}}{S}\,\delta\theta_o(t) + \frac{\theta_i}{S}\,\delta q_i(t) + \frac{1}{cS}\,\delta\nu(t)$$

Using the notation $V^* \equiv Sh^*$, $q_o^* \equiv \alpha\sqrt{(h^*)}$ and $\rho^* \equiv V^*/q_o^*$ (hold-up time of the tank), these linearized equations can be summarized in vector form.

The state equation is

$$\begin{pmatrix} \delta\dot{h}(t) \\ \delta\dot{\theta}_o(t) \end{pmatrix} = \begin{pmatrix} -1/2\rho & 0 \\ 0 & -1/\rho \end{pmatrix} \begin{pmatrix} \delta h(t) \\ \delta\theta_o(t) \end{pmatrix} + \begin{pmatrix} 1/S & 0 \\ (\theta_i - \theta_o^*)/V^* & 1/cV^* \end{pmatrix} \begin{pmatrix} \delta q_i(t) \\ \delta\nu(t) \end{pmatrix} \tag{1.37a}$$

and the output equation is

$$\begin{pmatrix} \delta q_o(t) \\ \delta\theta_o(t) \end{pmatrix} = \begin{pmatrix} q_o^*/2h^* & 0 \\ 0 & 1 \end{pmatrix} \begin{pmatrix} \delta h(t) \\ \delta\theta_o(t) \end{pmatrix} \tag{1.37b}$$

Example 1.5 (Model for stabilization of inverted pendulum). We consider the inverted pendulum shown in Figure 1.6. The pivot of the pendulum is mounted on a carriage which can move in a horizontal direction. The pendulum can be kept balanced at a specified position by applying horizontal forces to drive the carriage.

From inspection of Figure 1.6 we construct the differential equations describing the dynamics of the inverted pendulum and the carriage. The horizontal displacement of the pivot on the carriage is $\xi(t)$, while the rotational angle of the pendulum is $\phi(t)$. The horizontal and vertical positions of the centre of gravity of the

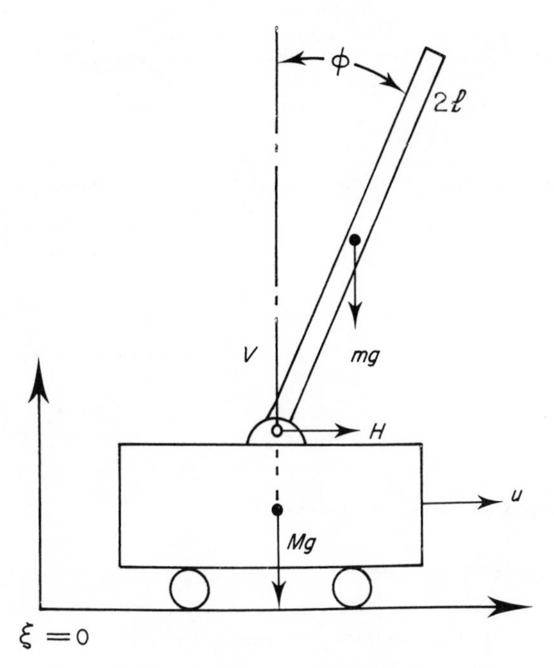

Figure 1.6 Inverted pendulum

pendulum are given by $\xi + l \sin \phi$ and $l \cos \phi$ respectively. The equations of motion can be seen to be

$$J \frac{d^2}{dt^2} \phi + C \frac{d}{dt} \phi = Vl \sin \phi - Hl \cos \phi$$

$$m \frac{d^2}{dt^2} (l \cos \phi) = V - mg$$

$$m \frac{d^2}{dt^2} (\xi + l \sin \phi) = H$$

$$M \frac{d^2}{dt^2} \xi + F \frac{d}{dt} \xi = u - H$$

where the length of the pendulum is $2l$, the moment of inertia with respect to the centre of gravity is $J = ml^3/3$, and H and V are horizontal and vertical reaction forces, respectively, on the pivot. C and F represent the friction coefficients for the rotary motion of the pendulum and the linear motion of the carriage, and u is the force driving the carriage.

Eliminating H and V from these equations gives

$$\begin{aligned} (J + ml^2)\ddot{\phi} + (ml \cos \phi)\ddot{\xi} &= -C\dot{\phi} + mlg \sin \phi \\ (M + m)\ddot{\xi} + (ml \cos \phi)\ddot{\phi} &= -F\dot{\xi} + (ml \sin \phi)\dot{\phi}^2 + u \end{aligned} \tag{1.38}$$

Using $\sin\phi \cong \phi$ and $\cos\phi \cong 1$ for small ϕ, we can derive the linearized equations with respect to $\delta\phi$ about $\phi^* = 0$ from (1.38). Then replacing $\delta\phi$ by ϕ gives

$$\begin{aligned} ml\ddot{\xi} + (J + ml^2)\ddot{\phi} &= -C\dot{\phi} + mlg\,\phi \\ (M+m)\ddot{\xi} + ml\ddot{\phi} &= -F\dot{\xi} + u \end{aligned} \tag{1.39}$$

Choosing the states $x_1 \equiv \xi$, $x_2 \equiv \phi$, $x_3 \equiv \dot{\xi}$ and $x_4 \equiv \dot{\phi}$, we obtain for the linear state space model of the inverted pendulum

$$\begin{pmatrix} \dot{x}_1 \\ \dot{x}_2 \\ \dot{x}_3 \\ \dot{x}_4 \end{pmatrix} = \begin{pmatrix} 0 & 0 & 1 & 0 \\ 0 & 0 & 0 & 1 \\ 0 & -m^2l^2g/\Delta & -F(J+ml^2)/\Delta & mlc/\Delta \\ 0 & ml(M+m)g/\Delta & mlF/\Delta & -C(M+m)/\Delta \end{pmatrix} \begin{pmatrix} x_1 \\ x_2 \\ x_3 \\ x_4 \end{pmatrix} + \begin{pmatrix} 0 \\ 0 \\ (J+ml^2)/\Delta \\ -ml/\Delta \end{pmatrix} u \tag{1.40}$$

where $\Delta \equiv (M+m)J + Mml^2$.

1.2.2 Analogies in Physical Systems

An analogy can be formed between a mechanical system, an electric circuit, and other physical systems. These analogies are used in replacing elements of one type of system with analogous quantities of another type of system.

Table 1.1 summarizes analogies of physical quantities between mechanical systems and electric circuits. For instance, Figure 1.7 shows a dynamic translation system and the corresponding electric circuit.

Let $y_1(t)$ and $y_2(t)$ be displacements from the equilibrium position which exist when the force $f(t)$ is zero. Then, the dynamic equations for the motion of the two masses are

$$M_1\ddot{y}_1 + D_1\dot{y}_1 + K_1y_1 + K_2(y_1 - y_2) = 0 \tag{1.41a}$$

$$M_2\ddot{y}_2 + D_2\dot{y}_2 + K_2(y_2 - y_1) = f(t). \tag{1.41b}$$

Next, we obtain equations for the electric circuit, shown in Figure 1.7(b), which is analogous to the mechanical system. If we apply Kirchhoff's voltage law to loops ① and ② of the circuit, we obtain

$$L_1\ddot{q}_1 + R_1\dot{q}_1 + q_1/C_1 + (q_1 - q_2)/C_2 = 0 \quad \text{for loop ①} \tag{1.42a}$$

$$L_2\ddot{q}_2 + R_2\dot{q}_2 + (q_2 - q_1)/C_2 = v(t) \quad \text{for loop ②} \tag{1.42b}$$

where $q_1 = \int i_1\,dt$ and $q_2 = \int i_2\,dt$.

Thus we see that the mechanical system and electric circuit yield similar differential equations with the analogous quantities shown in columns one and three of Table 1.1. This analogue is often referred to as the force–voltage analogue, and that between columns one and four as the force–current analogue (see Problem P1.4).

Table 1.1 Analogy of mechanical systems and electric circuits

	Mechanical translation system	Mechanical rotational system	Electric circuit (a)	Electric circuit (b)
Variables	Force: f Velocity: v	Torque: T Angular velocity: ω	Voltage: v Current: i	Current: i Voltage: v
	Distance: y $(y = \int v\,\mathrm{d}t)$	Angle: θ $(\theta = \int \omega\,\mathrm{d}t)$	Electric charge: q $(q = \int i\,\mathrm{d}t)$	Magnetic flux: ϕ $(\phi = \int v\,\mathrm{d}t)$
Elements and functions	Mass: M $(f = M\dot{v})$	Moment of inertia: J $(T = J\dot{\omega})$	Inductance: L $\left(v = L\dfrac{\mathrm{d}}{\mathrm{d}t}(i)\right)$	Capacitance: C $(i = C\dot{v})$
	Damping constant: D $(f = Dv)$	Damping constant: D $(T = D\omega)$	Resistance: R $(v = Ri)$	Conductance: G $(i = Gv)$
	Stiffness: K $(f = K\int v\,\mathrm{d}t)$	Stiffness: K $(T = K\int \omega\,\mathrm{d}t)$	Capacitance: C $\left(v = \dfrac{1}{C}\int i\,\mathrm{d}t\right)$	Inductance: L $\left(i = \dfrac{1}{L}\int v\,\mathrm{d}t\right)$

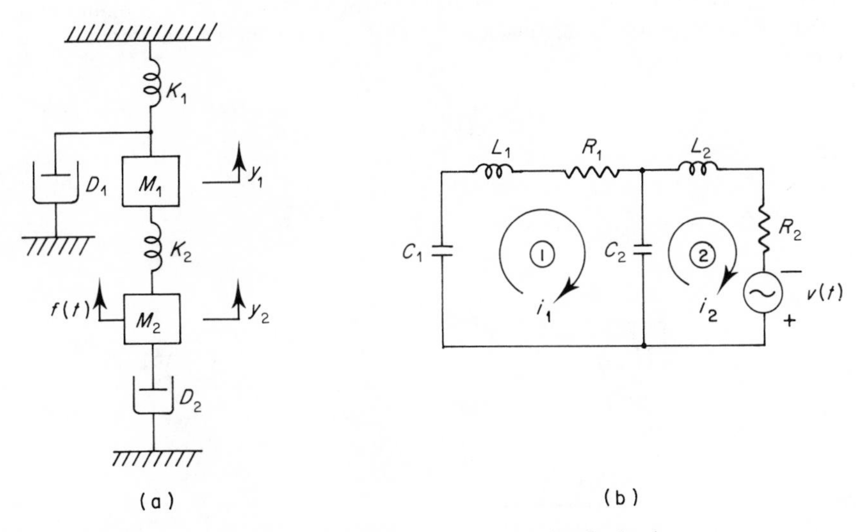

Figure 1.7 Mechanical system and electric circuit

1.2.3 Transfer Function and State Variable Description

We now consider the relationship between the transfer function and a state variable description by analysing a physical system.

Example 1.6 (Position control of d.c. motor). We consider the position control of an armature-controlled d.c. motor, shown in Figure 1.8, where the angular displacement of the motor shaft θ can be controlled by changing the armature voltage u. The torque T_m generated at the motor shaft depends on the magnetic field and the armature current i_a. In the absence of saturation T_m is linearly proportional to i_a, that is

$$T_m = K_T i_a \tag{1.43}$$

where K_T is the torque constant (Nm/A). The generated torque T_m drives the load and overcomes the viscous friction so that the equation of motion is

$$T_m = J_l \frac{d^2}{dt^2}\theta + D_l \frac{d}{dt}\theta \tag{1.44}$$

where J_l is the total moment of inertia of the motor shaft, including the inertia of the load, rotor and shaft, and D_l is the viscous friction constant.

For the armature circuit we have

$$L_a \frac{d}{dt} i_a + R_a i_a + V_a = u \tag{1.45}$$

where V_a is the back e.m.f. of the armature circuit, which is linearly proportional to the angular velocity of the motor, and given by

$$V_a = K_e \dot{\theta} \tag{1.46}$$

where K_e (V/rad/s) is the e.m.f. constant.

First, we give the transfer function representing the input–output behaviour of the d.c. motor. Taking the Laplace transforms of (1.43) to (1.46), assuming zero initial conditions, and eliminating i_a, V_a and T_m we obtain the transfer function from u to θ as

$$H(s) = \frac{\Theta(s)}{U(s)} = \frac{K_T}{s\{(L_a s + R_a)(J_l s + D_l) + K_e K_T\}} \tag{1.47}$$

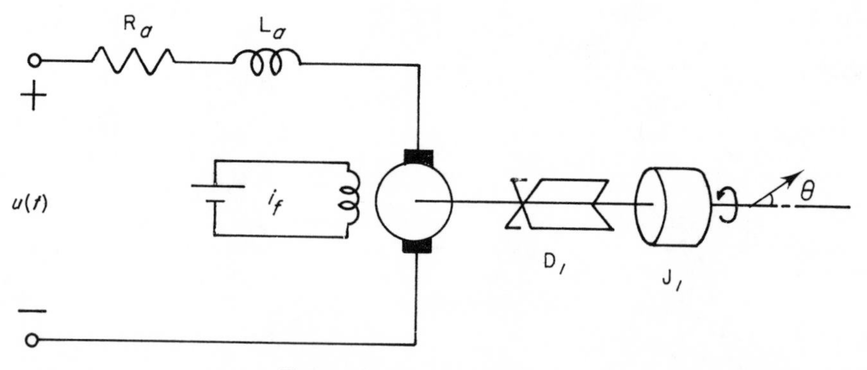

Figure 1.8 D.c. motor schematic

Next the state variable description of the motor is investigated. Let us choose the state variables as

$$x_1 = \theta, \quad x_2 = \dot{\theta}, \quad x_3 = \ddot{\theta} \tag{1.48}$$

Again eliminating i_a, V_a and T_m from (1.43) through (1.46) leads to the differential equation

$$\dddot{\theta} + \left(\frac{R_a}{L_a} + \frac{D_l}{J_l}\right)\ddot{\theta} + \frac{D_l R_a + K_e K_T}{J_l L_a}\dot{\theta} = \frac{K_T}{J_l L_a}u \tag{1.49}$$

It is noted from the choice of the state variables in (1.48) that

$$\begin{aligned} \dot{x}_1 &= \dot{\theta} = x_2 \\ \dot{x}_2 &= \ddot{\theta} = x_3 \\ \dot{x}_3 &= \dddot{\theta} \end{aligned} \tag{1.50}$$

so that we obtain the state space description

$$\begin{pmatrix} \dot{x}_1 \\ \dot{x}_2 \\ \dot{x}_3 \end{pmatrix} = \begin{pmatrix} 0 & 1 & 0 \\ 0 & 0 & 1 \\ 0 & -(R_a D_l + K_e K_T)/L_a J_l & -(R_a/L_a + D_l/J_l) \end{pmatrix} \begin{pmatrix} x_1 \\ x_2 \\ x_3 \end{pmatrix} + \begin{pmatrix} 0 \\ 0 \\ K_T/J_l L_a \end{pmatrix} u \tag{1.51a}$$

$$y = (1 \quad 0 \quad 0)\mathbf{x} \tag{1.51b}$$

As mentioned before, the choice of the state variables is not unique, therefore we can obtain other state variable descriptions. If, alternatively, the states are taken as

$$\bar{x}_1 = i_a, \quad \bar{x}_2 = \theta, \quad \bar{x}_3 = \dot{\theta}, \tag{1.52}$$

it is easy to see from (1.43) through (1.46) and (1.52) that

$$R_a \bar{x}_1 + L_a \dot{\bar{x}}_1 + K_e \bar{x}_3 = u$$

$$\dot{\bar{x}}_2 = \bar{x}_3$$

$$K_T \bar{x}_1 = J_l \dot{\bar{x}}_3 + D_l \bar{x}_3$$

and then the state variable description has the different form:

$$\begin{pmatrix} \dot{\bar{x}}_1 \\ \dot{\bar{x}}_2 \\ \dot{\bar{x}}_3 \end{pmatrix} = \begin{pmatrix} -R_a/L_a & 0 & -K_e/L_a \\ 0 & 0 & 1 \\ K_T/J_l & 0 & -D_l/J_l \end{pmatrix} \begin{pmatrix} \bar{x}_1 \\ \bar{x}_2 \\ \bar{x}_3 \end{pmatrix} + \begin{pmatrix} 1/L_a \\ 0 \\ 0 \end{pmatrix} u \tag{1.53a}$$

$$y = (0 \quad 1 \quad 0)\,\bar{\mathbf{x}} \tag{1.53b}$$

Calculating the transfer function for (1.53) using (1.22), we find, as expected, that the two state-variable descriptions (1.51) and (1.53) yield the same transfer function (1.47). The following relationships exist between the states $\mathbf{x}$ and $\bar{\mathbf{x}}$

$$x_1 = \bar{x}_2, \quad x_2 = \bar{x}_3, \quad K_T \bar{x}_1 = J_l x_3 + D_l x_2$$

or equivalently in vector form

$$\begin{pmatrix} x_1 \\ x_2 \\ x_3 \end{pmatrix} = \begin{pmatrix} 0 & 1 & 0 \\ 0 & 0 & 1 \\ K_T/J_l & 0 & -D_l/J_l \end{pmatrix} \begin{pmatrix} \bar{x}_1 \\ \bar{x}_2 \\ \bar{x}_3 \end{pmatrix} \equiv T\bar{\mathbf{x}} \tag{1.54}$$

This indicates the non-uniqueness of a state variable description of a given system and shows that an alternative set of state variables can be obtained by any non-singular transformation, T (det $T \neq 0$).

1.2.4 State Variable Description of an RLC Network

We explain a standard procedure for deriving the state equations of linear RLC networks through a typical example.

Example 1.7 Consider the linear electric circuit shown in Figure 1.9. If we know the initial conditions at time t_0, such as the magnetic flux $\phi(t_0)$ of the inductor and the charges $q_1(t_0)$ and $q_2(t_0)$ of the capacitors, and also the source voltage $e_s(t)$ and the source current $i_s(t)$ for time $t \geqq t_0$, then the flux $\phi(t)$ and the charges $q_1(t)$ and $q_2(t)$ are uniquely determined. This implies that the fluxes and charges can be taken as the state variables which can uniquely specify the behaviour of the RLC circuit. Assigning the inductor current $i_L(t)$ and capacitor voltages $v_c(t)$ as the state variables, since they satisfy the relationships $\phi(t) = Li_L(t)$ and $q(t) = Cv_c(t)$, the procedure is as follows:

(1) Choose the inductor current $i_L(t)$ and capacitor voltages $v_{c_1}(t)$ and $v_{c_2}(t)$ as the states, then

$$\begin{aligned} -L\frac{d}{dt}i_L &= v_L \\ -C_1\dot{v}_{c_1} &= i_{c1} \\ -C_2\dot{v}_{c_2} &= i_{c2} \end{aligned} \tag{1.55}$$

(2) Describe $v_L(t)$, $i_{c_1}(t)$ and $i_{c_2}(t)$ on the RHS of (1.55) in terms of the states $i_L(t)$, $v_{c_1}(t)$ and $v_{c2}(t)$ and the inputs $e_s(t)$ and $i_s(t)$, and, for this purpose, modify Figure 1.9 into Figure 1.10 which shows the equivalent circuit comprising only

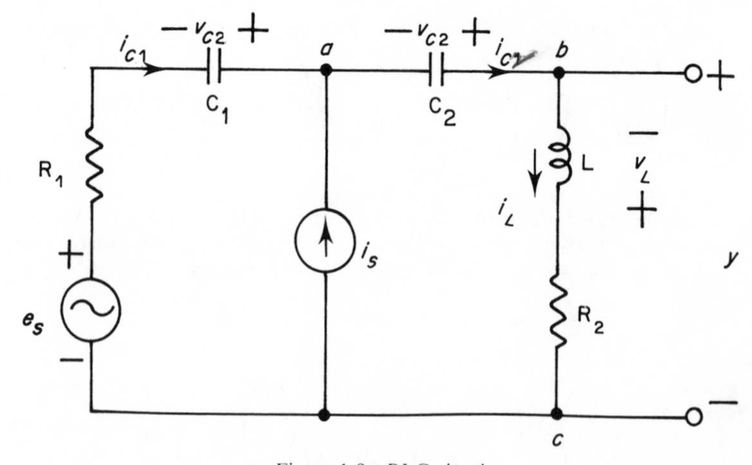

Figure 1.9 RLC circuit

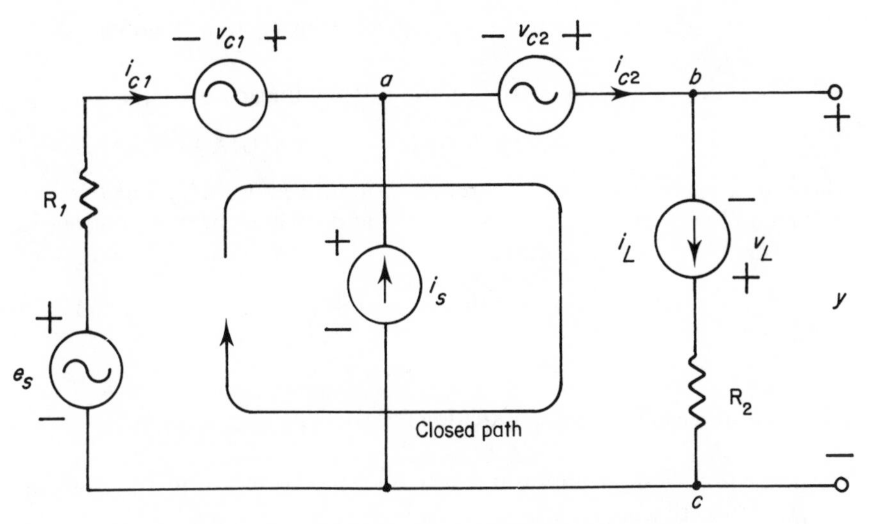

Figure 1.10 Equivalent circuit for use with Figure 1.9

resistors and sources. Now apply Kirchhoff's voltage and current laws to the circuit of Figure 1.10. The application of Kirchhoff's current law to the nodes a and b yields

$$i_{c_1}(t) + i_s(t) - i_{c_2}(t) = 0$$
$$i_{c_2}(t) - i_L(t) = 0$$

The application of Kirchhoff's voltage law to loop ① yields

$$e_s(t) - R_1 i_{c_1}(t) + v_{c_1}(t) + v_{c_2}(t) + v_L(t) - R_2 i_L(t) = 0$$

Solving these three equations with respect to $v_L(t)$, $i_{c_1}(t)$ and $i_{c2}(t)$ gives

$$\begin{aligned} i_{c_1}(t) &= i_L(t) - i_s(t) \\ i_{c_2}(t) &= i_L(t) \\ v_L(t) &= (R_1 + R_2) i_L(t) - v_{c_1}(t) - v_{c_2}(t) - e_s(t) - R_1 i_s(t) \end{aligned} \tag{1.56}$$

Finally substituting (1.56) into (1.55) and rearranging the equations into matrix form gives

$$\begin{pmatrix} \dot{i}_L(t) \\ \dot{v}_{c_1}(t) \\ \dot{v}_{c_2}(t) \end{pmatrix} = \begin{pmatrix} -(R_1 + R_2)/L & 1/L & 1/L \\ -1/C_1 & 0 & 0 \\ -1/C_2 & 0 & 0 \end{pmatrix} \begin{pmatrix} i_L(t) \\ v_{c_1}(t) \\ v_{c_2}(t) \end{pmatrix} + \begin{pmatrix} 1/L & R_1/L \\ 0 & 1/C_1 \\ 0 & 0 \end{pmatrix} \begin{pmatrix} e_s(t) \\ i_s(t) \end{pmatrix} \tag{1.57a}$$

$$\begin{aligned} y &= -v_L(t) + R_2 i_L(t) \\ &= (-R_1 \quad 1 \quad 1) \begin{pmatrix} i_L(t) \\ v_{c_1}(t) \\ v_{c_2}(t) \end{pmatrix} + (1 \quad R_1) \begin{pmatrix} e_s(t) \\ i_s(t) \end{pmatrix} \end{aligned} \tag{1.57b}$$

1.3 SOLUTION OF THE LINEAR STATE EQUATION

In this section we investigate the solution of the linear state equation

$$\dot{\mathbf{x}}(t) = A(t)\mathbf{x}(t) + B(t)\mathbf{u}(t), \qquad \text{for } \mathbf{x}(t_0) = \mathbf{x}_0 \tag{1.58}$$

where $\mathbf{x}(t)$ is an n-state vector, $\mathbf{u}(t)$ an m-input vector, $A(t)$ and $B(t)$ are $(n \times n)$ and $(n \times m)$ matrices respectively and each element is continuous with respect to t. We also discuss the time-invariant linear state equation

$$\dot{\mathbf{x}}(t) = A\mathbf{x}(t) + B\mathbf{u}(t), \qquad \text{for } \mathbf{x}(t_0) = \mathbf{x}_0. \tag{1.59}$$

1.3.1 Homogeneous Linear Equation

We first consider the solution of the homogeneous equation without any forcing terms, that is

$$\dot{\mathbf{x}}(t) = A(t)\mathbf{x}(t). \tag{1.60}$$

Let $\boldsymbol{\phi}_1(t, t_0), \boldsymbol{\phi}_2(t, t_0), \ldots, \boldsymbol{\phi}_n(t, t_0)$ be the set of solutions of (1.60) associated with the corresponding initial conditions

$$\mathbf{x}_1(t_0) = \mathbf{e}_1 = \begin{pmatrix} 1 \\ 0 \\ \vdots \\ 0 \end{pmatrix}, \mathbf{x}_2(t_0) = \mathbf{e}_2 = \begin{pmatrix} 0 \\ 1 \\ 0 \\ \vdots \\ 0 \end{pmatrix}, \ldots, \mathbf{x}_n(t_0) = \mathbf{e}_n = \begin{pmatrix} 0 \\ 0 \\ \vdots \\ 1 \end{pmatrix} \tag{1.61}$$

Combining these solutions, we define the $n \times n$ matrix

$$\Phi(t, t_0) = (\boldsymbol{\phi}_1(t, t_0), \boldsymbol{\phi}_2(t, t_0), \ldots, \boldsymbol{\phi}_n(t, t_0)) \tag{1.62}$$

which is called the *transition matrix.*

Properties of the transition matrix

(a) $\Phi(t, t_0)$ is the unique solution of the matrix differential equation for all t and t_0

$$\frac{\partial}{\partial t}\Phi(t, t_0) = A(t)\Phi(t, t_0) \qquad \text{for } t \geqq t_0 \tag{1.63}$$

$$\Phi(t_0, t_0) = I \tag{1.64}$$

Proof (1.63) and (1.64) are equivalent to the definition of the transition matrix. The uniqueness of the solution is assured by the continuity of $A(t)$.

(b) The homogenous equation (1.60) with the initial condition $\mathbf{x}(t_0) = \mathbf{x}_0$ has the solution

$$\mathbf{x}(t) = \Phi(t, t_0)\mathbf{x}_0 \qquad \text{for all } t \tag{1.65}$$

Proof Substituting (1.65) into (1.60), we have

$$\dot{\mathbf{x}}(t) = \frac{\partial}{\partial t}\Phi(t, t_0)\mathbf{x}_0 = A(t)\Phi(t, t_0)\mathbf{x}_0 = A(t)\mathbf{x}(t)$$

$$\mathbf{x}(t_0) = \Phi(t_0, t_0)\mathbf{x}_0 = \mathbf{x}_0.$$

(c) The transition matrix $\Phi(t, t_0)$ is non-singular.

Proof Suppose that $\Phi(t_1, t_0)$ is singular for some t_1. Then there exists a non-zero vector $\mathbf{c}$ ($\neq 0$) such that $\Phi(t_1, t_0)\mathbf{c} = 0$. The vector defined by $\psi(t) \equiv \Phi(t, t_0)\mathbf{c}$ is shown from (1.63) to satisfy $\dot{\psi}(t) = A(t)\psi(t)$ and $\psi(t_1) = 0$; hence it follows that $\psi(t) = 0$ for all t, i.e. $\Phi(t, t_0)\mathbf{c} = 0$ for all t. Since $\mathbf{c} \neq 0$, $\Phi(t, t_0)$ is singular for all t. This contradicts the non-singularity of $\Phi(t_0, t_0) = I$. Hence $\Phi(t, t_0)$ is non-singular for all t.

(d) For all t_0, t_1 and t_2,

$$\Phi(t_2, t_0) = \Phi(t_2, t_1)\Phi(t_1, t_0) \tag{1.66}$$

Proof Let the state variables at time t_0, t_1 and t_2 be denoted by $\mathbf{x}(t_0)$, $\mathbf{x}(t_1)$ and $\mathbf{x}(t_2)$, then

$$\mathbf{x}(t_1) = \Phi(t_1, t_0)\mathbf{x}(t_0) : \mathbf{x}(t_0) \to \mathbf{x}(t_1)$$
$$\mathbf{x}(t_2) = \Phi(t_2, t_1)\mathbf{x}(t_1) : \mathbf{x}(t_1) \to \mathbf{x}(t_2)$$
$$\mathbf{x}(t_2) = \Phi(t_2, t_0)\mathbf{x}(t_0) : \mathbf{x}(t_0) \to \mathbf{x}(t_2)$$

Hence we have

$$\mathbf{x}(t_2) = \Phi(t_2, t_1)\mathbf{x}(t_1) = \Phi(t_2, t_1)\Phi(t_1, t_0)\mathbf{x}(t_0).$$

Then from the uniqueness of the solution of (1.63), (1.66) can be established.

(e) For all t and t_0,

$$\Phi(t, t_0) = \Phi^{-1}(t_0, t)$$

Proof It is seen from (1.66) and (1.64) that

$$\Phi(t_0, t_0) = \Phi(t_0, t)\,\Phi(t, t_0) = I.$$

Hence for all t_0 and t, $\Phi(t, t_0) = \Phi^{-1}(t_0, t)$.

(f) The transition matrix $\Phi(t, t_0)$ can be expressed in the series

$$\Phi(t, t_0) = I + \int_{t_0}^{t} A(\tau)\,d\tau + \int_{t_0}^{t}\int_{t_0}^{\tau_1} A(\tau_1)A(\tau_2)\,d\tau_2\,d\tau_1 + \cdots$$

$$+ \int_{t_0}^{t}\int_{t_0}^{\tau_1}\cdots\int_{t_0}^{\tau_{k-1}} A(\tau_1)A(\tau_2)\ldots A(\tau_k)\,d\tau_k\ldots d\tau_1 + \cdots \tag{1.67}$$

Proof By substitution, it is easily verified that (1.67) is the solution of (1.63) with (1.64).

In the case where the matrix $A(t)$ is constant and equal to A, (1.67) becomes

$$\Phi(t, t_0) = I + A\int_{t_0}^{t} d\tau + A^2\int_{t_0}^{t}\int_{t_0}^{\tau_1} d\tau_2\,d\tau_1 + \cdots$$

$$= I + A(t - t_0) + \frac{1}{2!}A^2(t - t_0)^2 + \cdots \tag{1.68}$$

Hence the transition matrix depends only on $(t - t_0)$ and we can write

$$\Phi(t - t_0) \equiv \Phi(t - t_0, 0) \tag{1.69}$$

The series of (1.68) converges uniformly and absolutely on any finite interval, and extending the notation of the scalar exponential function to matrices we have

$$e^{At} = I + At + \frac{1}{2!}A^2t^2 + \cdots \tag{1.70}$$

so that the transition matrix of the time-invariant linear system can be written

$$\Phi(t) = e^{At} \tag{1.71}$$

Clearly this satisfies

$$\dot{\Phi}(t) = A\Phi(t), \qquad \Phi(0) = I \tag{1.72}$$

The series (1.70) is frequently utilized to numerically compute the transition matrix e^{At}.

1.3.2 Solution of the Inhomogeneous Linear Equation

The general solution of the inhomogeneous equation (1.58) is shown via an analogy with a scalar linear differential equation to consist of the sum of two parts: the first one being a particular solution of (1.58), which we now derive, and the second one being the general solution (1.65) of the homogeneous equation (1.60).

Theorem 1.1

The unique solution of the linear state equation

$$\dot{\mathbf{x}}(t) = A(t)\mathbf{x}(t) + B(t)\mathbf{u}(t) \qquad \text{for } \mathbf{x}(t_0) = \mathbf{x}_0 \tag{1.58}$$

is given by

$$\begin{aligned}\mathbf{x}(t) &= \psi(t; \mathbf{x}_0, t_0, \mathbf{u}) \\ &= \Phi(t, t_0)\mathbf{x}_0 + \int_{t_0}^{t} \Phi(t, \tau)B(\tau)\mathbf{u}(\tau)\, \mathrm{d}\tau \end{aligned}\tag{1.73}$$

where $\Phi(t, \tau)$ is the transition matrix satisfying (1.63) and (1.64).

Proof First, by setting t to t_0 in (1.73) it is easily seen that the solution (1.73) satisfies the initial condition $\mathbf{x}(t_0) = \mathbf{x}_0$. Next, we have to verify that (1.73) satisfies (1.58). Taking a derivative of (1.73) with respect to t, we obtain

$$\begin{aligned}\dot{\mathbf{x}}(t) &= \frac{\partial}{\partial t}\Phi(t, t_0)\mathbf{x}_0 + \frac{\partial}{\partial t}\int_{t_0}^{t} \Phi(t, \tau)B(\tau)\mathbf{u}(\tau)\, \mathrm{d}\tau \\ &= A(t)\Phi(t, t_0)\mathbf{x} + \Phi(t, t)B(t)\mathbf{u}(t) + \int_{t_0}^{t} \frac{\partial}{\partial t}\Phi(t, \tau)B(\tau)\mathbf{u}(\tau)\, \mathrm{d}\tau \\ &= A(t)\left[\Phi(t, t_0)\mathbf{x}_0 + \int_{t_0}^{t} \Phi(t, \tau)B(\tau)\mathbf{u}(\tau)\, \mathrm{d}\tau\right] + B(t)\mathbf{u}(t) \\ &= A(t)\mathbf{x}(t) + B(t)\mathbf{u}(t).\end{aligned}$$

Therefore, it is established that (1.73) is the unique solution of (1.58).

The solution (1.73) consists of the two terms. The first term expresses the solution of the homogeneous equation with $\mathbf{u}(t) = 0$ for all $t \geqq t_0$ as

$$\psi(t; \mathbf{x}_0, t_0, \mathbf{0}) = \Phi(t, t_0)\mathbf{x}_0 \tag{1.74}$$

and is the effect of the initial state $\mathbf{x}_0$ on the state at time t. The second term in (1.73) expresses the solution of (1.58) for the case of the zero initial state $\mathbf{x}_0 = 0$ as

$$\psi(t; \mathbf{0}, t_0, \mathbf{u}_{[t_0, t]}) = \int_{t_0}^{t} \Phi(t, \tau)B(\tau)\mathbf{u}(\tau)\, \mathrm{d}\tau \tag{1.75}$$

and is the effect of the input $\mathbf{u}_{[t_0, t]}$ on the state at time t.

The linear system has the property that the response of the state comprises the zero-input response and the zero-state response. $\psi(\cdot)$ satisfies the following properties:

(a) Decomposition property: for any t, t_0, $\mathbf{x}_0$ and $\mathbf{u}$,

$$\psi(t; \mathbf{x}_0, t_0, \mathbf{u}_{[t_0, t]}) = \psi(t; \mathbf{x}_0, t_0, \mathbf{0}) + \psi(t; \mathbf{0}, t_0, \mathbf{u}_{[t_0, t]})$$

(b) Linearity of the zero-state response: for any $t_0, t, \mathbf{u}^1, \mathbf{u}^2, \alpha_1, \alpha_2$,

$$\boldsymbol{\psi}(t; \mathbf{0}, t_0, \alpha_1 \mathbf{u}^1_{[t_0, t]} + \alpha_2 \mathbf{u}^2_{[t_0, t]}) = \alpha_1 \boldsymbol{\psi}(t; \mathbf{0}, t_0, \mathbf{u}^1_{[t_0, t]}) + \alpha_2 \boldsymbol{\psi}(t; \mathbf{0}, t_0, \mathbf{u}^2_{[t_0, t]})$$

(c) Linearity of the zero-input response: for any $t_0, t, \mathbf{x}^1, \mathbf{x}^2, \alpha_1, \alpha_2$,

$$\boldsymbol{\psi}(t; \alpha_1 \mathbf{x}^1 + \alpha_2 \mathbf{x}^2, t_0, \mathbf{0}) = \alpha_1 \boldsymbol{\psi}(t; \mathbf{x}^1, t_0, \mathbf{0}) + \alpha_2 \boldsymbol{\psi}(t; \mathbf{x}^2, t_0, \mathbf{0})$$

1.3.3 Calculation of the Transition Matrix

We now present several procedures for calculating the transition matrix e^{At} for a time-invariant linear system.

(a) Method via Laplace transform

Application of the Laplace transform $\mathscr{L}\{\cdot\}$ to (1.72) yields

$$s\mathscr{L}\{\Phi(t)\} - \Phi(0) = A\mathscr{L}\{\Phi(t)\}$$

hence

$$\Phi(t) = \mathscr{L}^{-1}\{(sI - A)^{-1}\} \tag{1.76}$$

A recursive scheme for computing $(sI - A)^{-1}$ was given in Section 1.1.3.

Example 1.8 Consider the linear system

$$\dot{\mathbf{x}} = \begin{pmatrix} 0 & -2 \\ 1 & -3 \end{pmatrix} \mathbf{x} + \begin{pmatrix} 1 \\ 1 \end{pmatrix} u, \qquad \text{for } \mathbf{x}(0) = \mathbf{x}_0$$

The solution is given as follows:

$$(sI - A)^{-1} = \left[\begin{pmatrix} s & 0 \\ 0 & s \end{pmatrix} - \begin{pmatrix} 0 & -2 \\ 1 & -3 \end{pmatrix} \right]^{-1} = \begin{pmatrix} s & 2 \\ -1 & s+3 \end{pmatrix}^{-1}$$

$$= \frac{1}{(s+1)(s+2)} \begin{pmatrix} s+3 & -2 \\ 1 & s \end{pmatrix}$$

$$= \begin{pmatrix} \dfrac{2}{s+1} - \dfrac{1}{s+2} & -\dfrac{2}{s+1} + \dfrac{2}{s+2} \\[2ex] \dfrac{1}{s+1} - \dfrac{1}{s+2} & -\dfrac{1}{s+1} + \dfrac{2}{s+2} \end{pmatrix}$$

Hence the transition matrix is

$$\Phi(t) = e^{At} = \mathscr{L}^{-1}\{(sI - A)^{-1}\} = \begin{pmatrix} 2e^{-t} - e^{-2t} & -2e^{-t} + 2e^{-2t} \\ e^{-t} - e^{-2t} & -e^{-t} + 2e^{-2t} \end{pmatrix}$$

and the general solution of the system equation is given from (1.73) by

$$\mathbf{x}(t) = \begin{pmatrix} 2e^{-t} - e^{-2t} & -2e^{-t} + 2e^{-2t} \\ e^{-t} - e^{-2t} & -e^{-t} + 2e^{-2t} \end{pmatrix} \mathbf{x}_0 + \int_{t_0}^{t} \begin{pmatrix} e^{-2(t-\tau)} \\ e^{-2(t-\tau)} \end{pmatrix} u(\tau)\, d\tau$$

(b) Method via diagonalization

Let $\lambda_1, \lambda_2, \ldots, \lambda_n$ be the distinct eigenvalues of A and let $\mathbf{v}_i$ be an eigenvector of A associated with λ_i for $i = 1, 2, \ldots, n$, then

$$A\mathbf{v}_i = \lambda_i \mathbf{v}_i \qquad \text{for } i = 1, 2, \ldots, n \tag{1.77}$$

The necessary and sufficient condition for (1.77) to have the non-zero solutions $\mathbf{v}_i \neq 0$ for $i = 1, \ldots, n$ is that λ_i is a root of the characteristic equation

$$\det(A - \lambda I) = 0 \tag{1.78}$$

The eigenvalues λ_i for $i = 1, \ldots, n$ are the solutions of (1.78), and are assumed to be all distinct, in which case the eigenvectors $\mathbf{v}_1, \ldots, \mathbf{v}_n$ are linearly independent. This is proved by contradiction. Suppose that $\mathbf{v}_1, \ldots, \mathbf{v}_n$ are linearly dependent, then there exists at least one non-zero $c_i \geqq 0$ such that

$$c_1\mathbf{v}_1 + c_2\mathbf{v}_2 + \cdots + c_n\mathbf{v}_n = 0$$

For instance, if $c_1 \neq 0$, we obtain from the above equation

$$(A - \lambda_2 I)(A - \lambda_3 I) \ldots (A - \lambda_n I)\left(\sum_{i=1}^{n} c_i\mathbf{v}_i\right) = \mathbf{0} \tag{1.79}$$

Since it follows clearly from (1.77) that

$$(A - \lambda_j I)\mathbf{v}_i = (\lambda_i - \lambda_j)\mathbf{v}_i,$$

then (1.79) can be reduced to

$$c_1(\lambda_1 - \lambda_2)(\lambda_1 - \lambda_3) \ldots (\lambda_1 - \lambda_n)\mathbf{v}_1 = \mathbf{0}.$$

By assumption that $\lambda_1, \ldots, \lambda_n$ are all distinct, we have $c_1 = 0$. This is a contradiction.

Therefore, since the eigenvectors are linearly independent, the matrix T defined by

$$T = (\mathbf{v}_1, \mathbf{v}_2, \ldots, \mathbf{v}_n) \tag{1.80}$$

is non-singular. Combination of (1.77) and (1.80) gives

$$AT = T\begin{pmatrix} \lambda_1 & 0 & \ldots & 0 \\ 0 & \lambda_2 & \ldots & 0 \\ 0 & 0 & & \vdots \\ & & \ldots & \\ 0 & & \ldots & \lambda_n \end{pmatrix} = T\Lambda \tag{1.81}$$

where Λ is a diagonal matrix with the eigenvalues in the diagonal elements. Hence we have

$$T^{-1}AT = \Lambda \tag{1.82}$$

We call the above operation diagonalization of the matrix A.

The transition matrix e^{At} can be calculated by use of the diagonalized matrix Λ of A, as

$$\Phi(t) \equiv e^{At} = T e^{\Lambda t} T^{-1} \tag{1.83}$$

This can easily verified by use of (1.70) as follows

$$\begin{aligned} T e^{\Lambda t} T^{-1} &= T e^{T^{-1}ATt} T^{-1} \\ &= T\left\{I + T^{-1}AT\,t + \frac{1}{2!}(T^{-1}AT)^2 t^2 + \cdots\right\}T^{-1} \\ &= TT^{-1} + TT^{-1}AT\,T^{-1}t + \frac{1}{2!}T(T^{-1}AT)^2T^{-1}t^2 + \cdots \\ &= I + At + \frac{1}{2!}A^2t^2 + \cdots = e^{At} \end{aligned} \tag{1.84}$$

Example 1.9 Calculate e^{At} for

$$A = \begin{pmatrix} 0 & 1 & 0 \\ 3 & 0 & 2 \\ -12 & -7 & -6 \end{pmatrix}.$$

From (1.78), $\det(A - \lambda I) = -(\lambda+1)(\lambda+2)(\lambda+3)$, then $\lambda_1 = -1, \lambda_2 = -2$ and $\lambda_3 = -3$. The matrix T associated with the eigenvalues is

$$T = \begin{pmatrix} 1 & 2 & 1 \\ -1 & -4 & -3 \\ -1 & 1 & 3 \end{pmatrix}, \text{ and } T^{-1} = \begin{pmatrix} 9/2 & 5/2 & 1 \\ -3 & -2 & -1 \\ 5/2 & 3/2 & 1 \end{pmatrix}$$

hence

$$\Lambda = T^{-1}AT = \begin{pmatrix} -1 & 0 & 0 \\ 0 & -2 & 0 \\ 0 & 0 & -3 \end{pmatrix}.$$

From (1.83), we have

$$e^{At} = \begin{pmatrix} 1 & 2 & 1 \\ -1 & -4 & -3 \\ -1 & 1 & 3 \end{pmatrix} \begin{pmatrix} e^{-t} & 0 & 0 \\ 0 & e^{-2t} & 0 \\ 0 & 0 & e^{-3t} \end{pmatrix} \begin{pmatrix} 9/2 & 5/2 & 1 \\ -3 & -2 & -1 \\ 5/2 & 3/2 & 1 \end{pmatrix}$$

$$= \begin{pmatrix} \frac{9}{2}e^{-t} - 6e^{-2t} + \frac{5}{2}e^{-3t} & \frac{5}{2}e^{-t} - 4e^{-2t} + \frac{3}{2}e^{-3t} & e^{-t} - 2e^{-2t} + e^{-3t} \\ -\frac{9}{2}e^{-t} + 12e^{-2t} - \frac{15}{2}e^{-3t} & -\frac{5}{2}e^{-t} + 8e^{-2t} - \frac{9}{2}e^{-3t} & -e^{-t} + 4e^{-2t} - 3e^{-3t} \\ -\frac{9}{2}e^{-t} - 3e^{-2t} + \frac{15}{2}e^{-3t} & -\frac{5}{2}e^{-t} - 2e^{-2t} + \frac{9}{2}e^{-3t} & -e^{-t} - e^{-2t} + 3e^{-3t} \end{pmatrix}$$

If the matrix A has repeated eigenvalues, it is not always possible to find a diagonalized matrix like (1.81). In this case, the matrix Λ becomes of the Jordan block form which is discussed further in Section 1.5.3.

(c) Sylvester expansion theorem

Let $\lambda_1, \lambda_2, \ldots, \lambda_n$ be the distinct eigenvalues of A, and let $f(A)$ be a matrix polynomial of degree $n-1$. Then the Sylvester theorem states that $f(A)$ can be expressed as

$$f(A)=\sum_{i=1}^{n} f(\lambda_i)\frac{(A-\lambda_1 I)\ldots(A-\lambda_{i-1}I)(A-\lambda_{i+1}I)\ldots(A-\lambda_n I)}{(\lambda_i-\lambda_1)\ldots(\lambda_i-\lambda_{i-1})(\lambda_i-\lambda_{i+1})\ldots(\lambda_i-\lambda_n)} \tag{1.85}$$

The case of $f(A)=\mathrm{e}^{At}$ corresponds to the calculation of the transition matrix. We use an example to illustrate the procedure.

Example 1.10 Calculate

$$f(A)=\mathrm{e}^{At} \text{ with } A=\begin{pmatrix}-1 & 1\\ 0 & -2\end{pmatrix}.$$

From (1.78), $\det(\lambda I - A)=(\lambda+1)(\lambda+2)=0$, thus $\lambda_1=-1$ and $\lambda_2=-2$. Hence $f(\lambda_1)=\mathrm{e}^{-t}$, $f(\lambda_2)=\mathrm{e}^{-2t}$ and thus from (1.85) we have

$$\begin{aligned}\mathrm{e}^{At} &= f(\lambda_1)\frac{A-\lambda_2 I}{\lambda_1-\lambda_2}+f(\lambda_2)\frac{A-\lambda_1 I}{\lambda_2-\lambda_1}\\ &= \mathrm{e}^{-t}(A+2I)+\mathrm{e}^{-2t}(-A-I)=\begin{pmatrix}\mathrm{e}^{-t} & \mathrm{e}^{-t}-\mathrm{e}^{-2t}\\ 0 & \mathrm{e}^{-2t}\end{pmatrix}\end{aligned}$$

(d) Use of the Cayley-Hamilton theorem

We have already seen from the Cayley-Hamilton theorem that for any integer i an $n\times n$ matrix A^{n+i} can be expressed as a linear combination of $I, A, \ldots, A^{n-1}$. Thus e^{At} can be written

$$\mathrm{e}^{At}=\alpha_0 I+\alpha_1 A+\alpha_2 A^2+\cdots+\alpha_{n-1}A^{n-1}.$$

In addition if A has distinct eigenvalues they must satisfy the same equation, that is

$$\mathrm{e}^{\lambda_i t}=\alpha_0+\alpha_1\lambda_i+\alpha_2\lambda_i^2+\cdots+\alpha_{n-1}\lambda_i^{n-1}$$

for $i=1,2,\ldots,n$. These n equations then enable the n unknown coefficients $\alpha_0, \alpha_1, \ldots, \alpha_{n-1}$, to be found which are required to evaluate e^{At}.

1.4 EQUIVALENT SYSTEMS

1.4.1 Input–Output Relation

We consider the linear system

$$\dot{\mathbf{x}}(t) = A(t)\mathbf{x}(t) + B(t)\mathbf{u}(t) \tag{1.86a}$$

$$\mathbf{y}(t) = C(t)\mathbf{x}(t) + D(t)\mathbf{u}(t) \tag{1.86b}$$

where $\mathbf{x}(t)$, $\mathbf{u}(t)$ and $\mathbf{y}(t)$ are an n-state vector, an m-input vector and a p-output vector respectively; $A(t)$, $B(t)$, $C(t)$, and $D(t)$ are an $n \times n$, $n \times m$, $p \times n$, and $p \times m$ matrix respectively; and all of them are assumed to be continuous in t.

The general solution of the state equation (1.86a) is known to be

$$\begin{aligned}\mathbf{x}(t) &= \psi(t; \mathbf{x}_0, t_0, \mathbf{u}_{[t_0, t]}) \\ &= \Phi(t, t_0)\mathbf{x}_0 + \int_{t_0}^{t} \Phi(t, \tau)B(\tau)\mathbf{u}(\tau)\,\mathrm{d}\tau \end{aligned} \tag{1.87}$$

Then, it follows from (1.86) that the output $\mathbf{y}(t)$ is

$$\mathbf{y}(t) = \mathbf{y}(t; \mathbf{x}_0, t_0, \mathbf{u}_{[t_0, t]})$$

$$= C(t)\Phi(t, t_0)\mathbf{x}_0 + C(t)\int_{t_0}^{t} \Phi(t, \tau)B(\tau)\mathbf{u}(\tau)\,\mathrm{d}\tau + D(t)\mathbf{u}(t) \tag{1.88}$$

$$= \mathbf{y}(t; \mathbf{x}_0, t_0, \mathbf{0}) + y(t; \mathbf{0}, t_0, \mathbf{u}_{[t_0, t]}) \tag{1.89}$$

Thus the output $\mathbf{y}(t)$ can be decomposed into the zero-input response and the zero-state response, as indicated by (1.89). The first term in (1.88), that is the zero-input response, represents the effect of the initial state $\mathbf{x}_0$ on the output response $\mathbf{y}(t)$. The second and third terms in (1.88), that is the zero-state response, express the output response in the case of a zero initial state, and can be written as

$$\mathbf{y}(t; \mathbf{0}, t_0, \mathbf{u}_{[t_0, t]}) = \int_{t_0}^{t} \{C(t)\Phi(t, \tau)B(\tau) + D(t)\delta^+(t-\tau)\}\mathbf{u}(\tau)\,\mathrm{d}\tau \tag{1.90}$$

where $\delta^+(t)$ is the delta function defined by

$$\delta^+(t) = \lim_{\Delta \to 0} \delta_\Delta^+(t)$$

and

$$\delta_\Delta^+(t) = \begin{cases} 1/\Delta & \text{for } 0 \leqq t \leqq \Delta \\ 0 & \text{otherwise} \end{cases}$$

If $H(t, \tau)$ is denoted by

$$H(t, \tau) \equiv \begin{cases} C(t)\Phi(t, \tau)B(\tau) + D(t)\delta^+(t-\tau); & \text{for } \tau \leqq t \\ 0 & ; \text{for } \tau > t \end{cases} \tag{1.91}$$

the output $y(t)$ for the zero-initial state $\mathbf{x}_0 = 0$ can be described by

$$\mathbf{y}(t) = \mathbf{y}(t; \mathbf{0}, t_0, \mathbf{u}_{[t_0, \infty)}) = \int_{t_0}^{\infty} H(t, \tau)\mathbf{u}(\tau)\, \mathrm{d}\tau \tag{1.92}$$

Hence, it can be seen from (1.4) that $H(t, \tau)$ is the impulse response. We also remark that, if the state-variable description $[A(t), B(t), C(t), D(t)]$ is given, the impulse response function is uniquely determined as (1.91).

In particular for the time-invariant linear dynamical system (A, B, C, D)

$$\dot{\mathbf{x}}(t) = A\mathbf{x}(t) + B\mathbf{u}(t) \tag{1.93a}$$

$$\mathbf{y}(t) = C\mathbf{x}(t) + D\mathbf{u}(t) \tag{1.93b}$$

the transition matrix $\Phi(t, \tau)$ is given by $\Phi(t - \tau) = \mathrm{e}^{A(t-\tau)}$ and the matrix impulse response is

$$H(t, \tau) = C\,\mathrm{e}^{A(t-\tau)}B + D\delta^{+}(t - \tau) \qquad \text{for } t \geqq \tau.$$

Since $H(t, \tau)$ is a function of only $(t - \tau)$, then we can write it as

$$H(t) = C\,e^{At}B + D\delta^{+}(t) \qquad \text{for } t \geqq 0 \tag{1.94}$$

By taking the Laplace transform of (1.94), we obtain the transfer function $H(s)$ as

$$H(s) = C(sI - A)^{-1}B + D \tag{1.95}$$

which is the same as (1.22) obtained directly from the Laplace transform of (1.16).

1.4.2 Equivalent Systems

We consider a transformation of the state vectors of a linear dynamical system by use of the non-singular matrix $T(t)$, that is

$$\bar{\mathbf{x}}(t) = T^{-1}(t)\mathbf{x}(t) \quad \text{or} \quad \mathbf{x}(t) = T(t)\bar{\mathbf{x}}(t) \tag{1.96}$$

where $T(t)$ is non-singular for all t and continuously differentiable.

The new state $\bar{\mathbf{x}}(t)$ obtained through the transformation (1.96) satisfies

$$\dot{\bar{\mathbf{x}}}(t) = \bar{A}(t)\bar{\mathbf{x}}(t) + \bar{B}(t)\mathbf{u}(t) \tag{1.97a}$$

$$\mathbf{y}(t) = \bar{C}(t)\bar{\mathbf{x}}(t) + \bar{D}(t)\mathbf{u}(t) \tag{1.97b}$$

where

$$\bar{A}(t) = T^{-1}(t)A(t)T(t) - T^{-1}(t)\dot{T}(t) \tag{1.98a}$$

$$\bar{B}(t) = T^{-1}(t)B(t) \tag{1.98b}$$

$$\bar{C}(t) = C(t)T(t) \tag{1.98c}$$

$$\bar{D}(t) = D(t) \tag{1.98d}$$

We show below in Theorem 1.2 that the input–output relationship is not changed by the transformation. Thus, if the initial conditions also satisfy (1.96), that is, $\mathbf{x}(t_0) = T(t_0)\bar{\mathbf{x}}(t_0)$, then both linear systems have identical output responses for the same input. In this case the linear dynamical system given by (1.97) is said to be *equivalent* to the linear system of (1.86), and $T(t)$ is also called an *equivalence transformation*.

Similarly, dynamical systems which are equivalent to the time-invariant linear dynamical system in (1.93) can be derived through the non-singular transformation matrix T given by

$$\bar{\mathbf{x}}(t) = T^{-1}\mathbf{x}(t) \quad \text{or} \quad \mathbf{x}(t) = T\bar{\mathbf{x}}(t) \tag{1.99}$$

so that

$$\dot{\bar{\mathbf{x}}}(t) = \bar{A}\bar{\mathbf{x}}(t) + \bar{B}\mathbf{u}(t) \tag{1.100a}$$

$$\mathbf{y}(t) = \bar{C}\bar{\mathbf{x}}(t) + \bar{D}\mathbf{u}(t) \tag{1.100b}$$

where

$$\bar{A} = T^{-1}AT \tag{1.101a}$$

$$\bar{B} = T^{-1}B \tag{1.101b}$$

$$\bar{C} = CT \tag{1.101c}$$

$$\bar{D} = D \tag{1.101d}$$

Hence, there exist an infinite number of equivalent systems since the transformation matrix can be arbitrarily chosen.

Theorem 1.2

Equivalent linear systems have identical matrix impulse responses.

Proof The state variables $\mathbf{x}(t)$ and $\bar{\mathbf{x}}(t)$ for the linear dynamical systems described by (1.86) and (1.97), respectively, are given by

$$\mathbf{x}(t) = \Phi(t, t_0)\mathbf{x}(t_0) + \int_{t_0}^{t} \Phi(t, \tau)B(\tau)\mathbf{u}(\tau)\,d\tau \tag{1.102}$$

$$\bar{\mathbf{x}}(t) = \bar{\Phi}(t, t_0)\bar{\mathbf{x}}(t_0) + \int_{t_0}^{t} \bar{\Phi}(t, \tau)\bar{B}(\tau)\mathbf{u}(\tau)\,d\tau \tag{1.103}$$

where $\Phi(t, \tau)$ and $\bar{\Phi}(t, \tau)$ are the transition matrices corresponding to $A(t)$ and $\bar{A}(t)$ respectively. Substituting (1.96) into (1.102) and multiplying by $T^{-1}(t)$, we have

$$\bar{\mathbf{x}}(t) = T^{-1}(t)\Phi(t, t_0)T(t_0)\bar{\mathbf{x}}(t_0) + \int_{t_0}^{t} T^{-1}(t)\Phi(t, \tau)T(\tau)T^{-1}(\tau)B(\tau)\mathbf{u}(\tau)\,d\tau \tag{1.104}$$

Then, it follows from (1.103) and (1.98) that

$$\bar{\Phi}(t,\tau) = T^{-1}(t)\Phi(t,\tau)T(\tau) \tag{1.105}$$

Using (1.105) and (1.98) in (1.91) it is seen that the impulse response $H(t, \tau)$ for the dynamical system (1.97) is

$$\begin{aligned}\bar{H}(t,\tau) &= \bar{C}(t)\bar{\Phi}(t,\tau)\bar{B}(\tau) + \bar{D}(t)\delta^+(t-\tau)\\ &= C(t)T(t)T^{-1}(t)\Phi(t,\tau)T(\tau)T^{-1}(\tau)B(\tau) + D(t)\delta^+(t-\tau)\\ &= C(t)\Phi(t,\tau)B(\tau) + D(t)\delta^+(t-\tau)\\ &= H(t,\tau)\end{aligned} \tag{1.106}$$

Thus, the theorem is established.

Corollary 1.1

Linear time-invariant dynamical systems which are equivalent have identical transfer functions.

Proof The transfer function associated with the linear system (1.100) is

$$\begin{aligned}\bar{H}(s) &= \bar{C}(sI-\bar{A})^{-1}\bar{B} + \bar{D} = CT(sT^{-1}T - T^{-1}AT)T^{-1}B + D\\ &= CTT^{-1}(sI-A)^{-1}TT^{-1}B + D = C(sI-A)^{-1}B + D\\ &= H(s)\end{aligned}$$

Hence all equivalent linear systems have identical transfer functions.

1.5 REALIZATION OF SINGLE-INPUT AND SINGLE-OUTPUT LINEAR SYSTEMS

We consider, initially, a linear multi-input multi-output (MIMO) system with transfer function matrix $H(s)$. If there exists a linear finite-dimensional dynamical system (A, B, C, D) that has the specified $H(s)$, the transfer function matrix $H(s)$ is said to be *realizable,* and (A, B, C, D) is called a *realization* of $H(s)$.

When the system description (A, B, C, D) is given, $H(s)$ is described by the rational function of s indicated in (1.23). The determinant of $(sI - A)$ gives a denominator polynomial of degree n in s, while every element of the adjoint matrix of $(sI - A)$ gives a numerator polynomial of degree equal to or less than $n-1$. If $D=0$ the transfer function matrix $H(s) = C(sI-A)^{-1}B$, and for each element the degree of its denominator is at least one degree higher than that of its numerator. In this case the transfer function is called *strictly proper*. If D is a constant matrix, the degree of the denominator of each element is equal to or greater than that of its numerator, and then the transfer function is called *proper*. Hence, it can be seen that $H(s)$ is realizable if and only if $H(s)$ is a proper rational matrix.

We now turn our attention to the realization of a single-input single-output rational transfer function.

$$H(s) = \frac{b_n s^n + b_{n-1}s^{n-1} + \cdots + b_1 s + b_0}{s^n + \alpha_{n-1}s^{n-1} + \cdots + \alpha_1 s + \alpha_0} \tag{1.107a}$$

$$= \frac{\beta_{n-1}s^{n-1} + \beta_{n-2}s^{n-2} + \cdots + \beta_1 s + \beta_0}{s^n + \alpha_{n-1}s^{n-1} + \cdots + \alpha_1 s + \alpha_0} + \delta \tag{1.107b}$$

where $\beta_i = b_i - b_n\alpha_i$ for $i = 0, 1, \ldots, n-1$ and $\delta = b_n$. We present methods for reconstructing state variable descriptions $(A, \mathbf{b}, \mathbf{c}^{\mathrm{T}}, d)$ for the given transfer function (1.107b).

1.5.1 Companion Form (a): Controllable canonical form

Equation (1.107b) being decomposed into two parts, as shown in Figure 1.11, the output $Y(s)$ can be written as

$$Y(s) = Z(s) + \delta U(s) \tag{1.108}$$

where

$$\frac{Z(s)}{U(s)} = \frac{\beta_{n-1}s^{n-1} + \beta_{n-2}s^{n-2} + \cdots + \beta_1 s + \beta_0}{s^n + \alpha_{n-1}s^{n-1} + \cdots + \alpha_1 s + \alpha_0} \tag{1.109}$$

Let $X_1(s)$ be defined as in Fig. 1.11, then we have

$$\frac{X_1(s)}{U(s)} = \frac{1}{s^n + \alpha_{n-1}s^{n-1} + \cdots + \alpha_1 s + \alpha_0} \tag{1.110}$$

$$\frac{Z(s)}{X_1(s)} = \beta_{n-1}s^{n-1} + \beta_{n-2}s^{n-2} + \cdots + \beta_1 s + \beta_0 \tag{1.111}$$

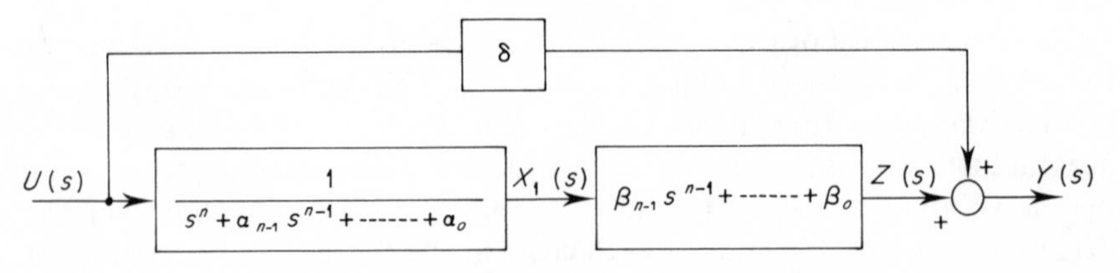

Figure 1.11 Transfer function representation

If $x_1(t) = \mathscr{L}^{-1}\{X_1(s)\}$ is taken as one of the state variables, it satisfies the linear ordinary differential equation

$$x_1^{(n)}(t) + \alpha_{n-1}x_1^{(n-1)}(t) + \cdots + \alpha_0 x_1(t) = u(t) \tag{1.112}$$

Further, if the other $n-1$ state variables are chosen as

$$x_2 = \dot{x}_1,\ x_3 = \dot{x}_2 = x_1^{(2)},\ x_4 = \dot{x}_3 = x_1^{(3)}, \ldots, x_n = \dot{x}_{n-1} = x_1^{(n-1)},$$

it follows from (1.112) that the state equation is

$$\begin{aligned} \dot{x}_1 &= x_2 \\ \dot{x}_2 &= x_3 \\ &\vdots \\ \dot{x}_{n-1} &= x_n \\ \dot{x}_n &= -\alpha_0 x_1 - \alpha_1 x_2 - \cdots - \alpha_{n-1} x_n + u \end{aligned} \tag{1.113}$$

and from (1.108) and (1.111) that the output equation is

$$\begin{aligned} y(t) &= z(t) + \delta u(t) \\ &= \beta_0 x_1(t) + \beta_1 x_2(t) + \cdots + \beta_{n-1} x_n(t) + \delta u(t) \end{aligned} \tag{1.114}$$

Rewriting (1.113) and (1.114) in matrix form, we have

$$\dot{\mathbf{x}}(t) = A\mathbf{x}(t) + \mathbf{b}u(t) \tag{1.115a}$$

$$y(t) = \mathbf{c}^{\mathrm{T}}\mathbf{x}(t) + du(t) \tag{1.115b}$$

where

$$A = \begin{pmatrix} 0 & 1 & 0 & & \cdots & 0 \\ 0 & 0 & 1 & & \cdots & 0 \\ 0 & 0 & 0 & 1 & \cdots & 0 \\ \vdots & & & & \ddots & \vdots \\ & & & & \cdots & 1 \\ -\alpha_0 & -\alpha_1 & \cdots & & -\alpha_{n-2} & -\alpha_{n-1} \end{pmatrix}, \quad \mathbf{b} = \begin{pmatrix} 0 \\ 0 \\ \vdots \\ 0 \\ 1 \end{pmatrix}$$

$$\mathbf{c}^{\mathrm{T}} = (\beta_0 \quad \beta_1 \quad \beta_2 \quad \ldots \quad \beta_{n-1}),\ d = \delta$$

The block diagram of the state space description (1.115) is given in Figure 1.12.

1.5.2 Companion Form (b): Observable canonical form

If the companion form $(A, \mathbf{b}, \mathbf{c}^{\mathrm{T}}, d)$ in (1.115) is a realization of $H(s)$, another companion form $(A^{\mathrm{T}}, \mathbf{c}, \mathbf{b}^{\mathrm{T}}, d)$ is also a realization of $H(s)$, since

$$\begin{aligned} H(s) = \mathbf{c}^{\mathrm{T}}(sI - A)^{-1}\mathbf{b} &= \{\mathbf{b}^{\mathrm{T}}(sI - A^{\mathrm{T}})^{-1}\mathbf{c}\}^{\mathrm{T}} \\ &= \mathbf{b}^{\mathrm{T}}(sI - A)^{-1}\mathbf{c}. \end{aligned} \tag{1.116}$$

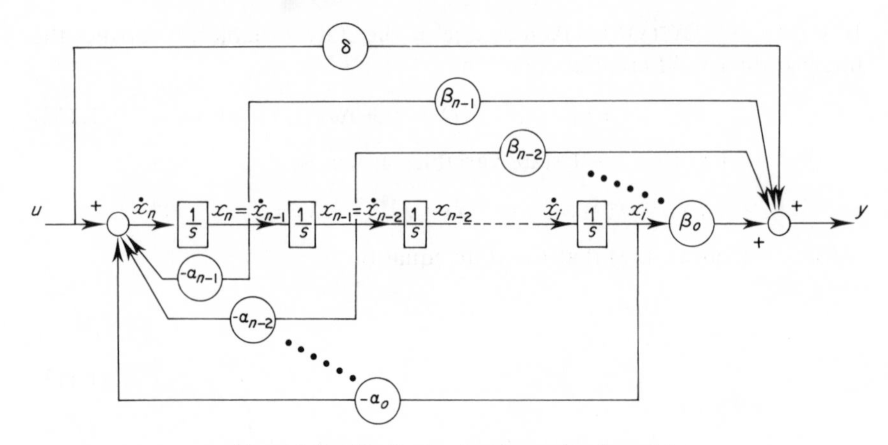

Figure 1.12 Realization by companion form (a)

Thus we can obtain another companion form given by

$$\begin{pmatrix} \dot{x}_1 \\ \dot{x}_2 \\ \dot{x}_3 \\ \vdots \\ \dot{x}_n \end{pmatrix} = \begin{pmatrix} 0 & 0 & & \cdots & & -\alpha_0 \\ 1 & 0 & & \cdots & & -\alpha_1 \\ 0 & 1 & 0 & \cdots & & -\alpha_2 \\ \vdots & \vdots & 1 & \vdots & & \vdots \\ 0 & \cdots & \vdots & 0 & 1 & -\alpha_{n-1} \end{pmatrix} \begin{pmatrix} x_1 \\ x_2 \\ x_3 \\ \vdots \\ x_n \end{pmatrix} + \begin{pmatrix} \beta_0 \\ \beta_1 \\ \beta_2 \\ \vdots \\ \beta_{n-1} \end{pmatrix} u \quad (1.117a)$$

$$y = (0 \quad 0 \quad \ldots \quad 0 \quad 1)\mathbf{x} + \delta u. \quad (1.117b)$$

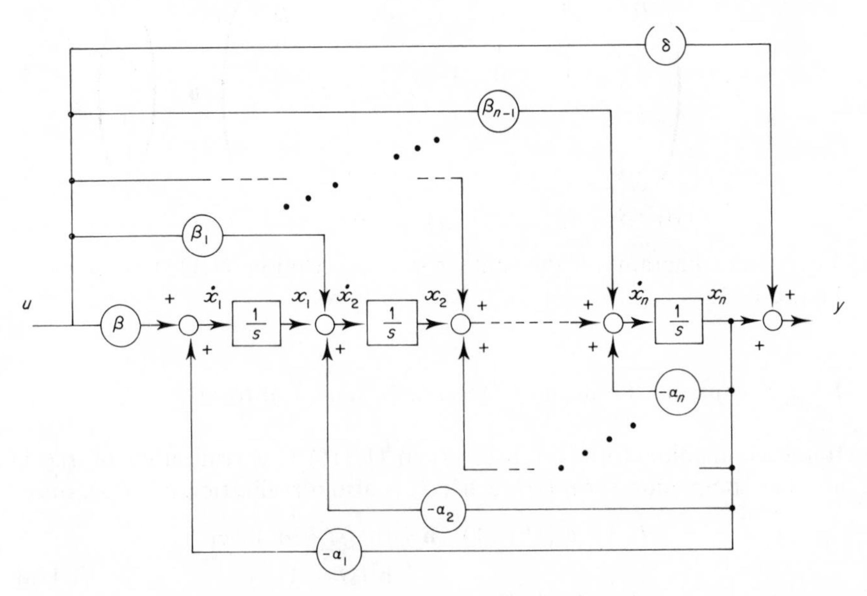

Figure 1.13 Realization by companion form (b)

It is seen, from Figure 1.13 which gives the block diagram of this realization, that the state variables are assigned by

$$\begin{aligned} x_{n-1} &= \dot{x}_n + \alpha_{n-1}x_n - \beta_{n-1}u \\ x_{n-2} &= \dot{x}_{n-1} + \alpha_{n-2}x_n - \beta_{n-2}u \\ &\cdots \\ x_1 &= \dot{x}_2 + \alpha_1 x_n - \beta_1 u \\ 0 &= \dot{x}_1 + \alpha_0 x_n - \beta_0 u. \end{aligned} \tag{1.118}$$

1.5.3 Jordan Canonical Form

When the transfer function (1.107) has distinct poles $\lambda_1, \lambda_2, \ldots, \lambda_n$, $H(s)$ can be described by the partial fraction expansion

$$H(s) = \frac{\gamma_1}{s-\lambda_1} + \frac{\gamma_2}{s-\lambda_2} + \cdots + \frac{\gamma_n}{s-\lambda_n} + \delta \tag{1.119}$$

where

$$\gamma_i = \lim_{s \to \lambda_i} (s - \lambda_i)H(s)$$

As illustrated in Figure 1.14, if we assign the state variables $x_1, x_2, \ldots, x_n$ as

$$\frac{X_1(s)}{U(s)} = \frac{1}{s-\lambda_1}, \quad \frac{X_2(s)}{U(s)} = \frac{1}{s-\lambda_2}, \quad \ldots, \quad \frac{X_n(s)}{U(s)} = \frac{1}{s-\lambda_n},$$

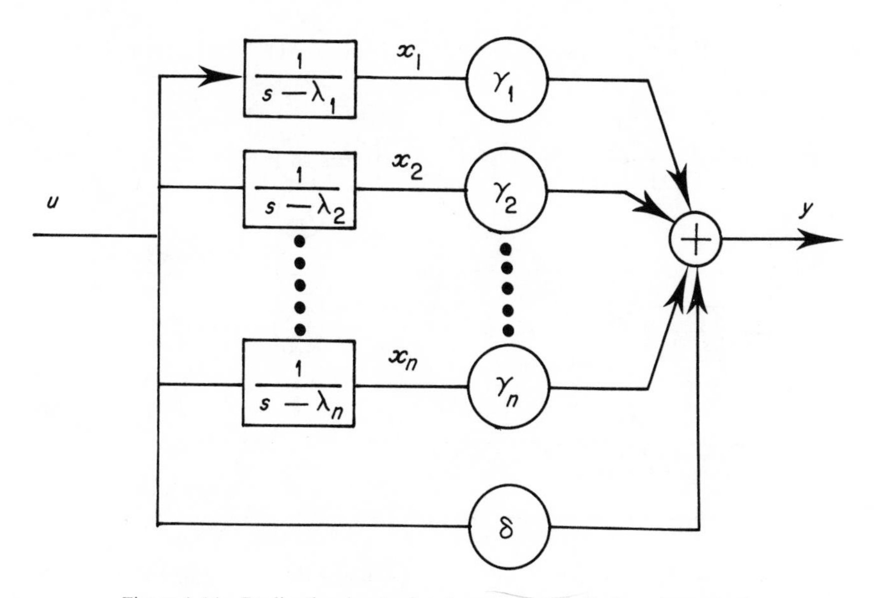

Figure 1.14 Realization by Jordan form: case (a) distinct eigenvalues

the output $Y(s)$ is

$$Y(s) = \gamma_1 X_1(s) + \gamma_2 X_2(s) + \cdots + \gamma_n X_n(s) + \delta U(s) \tag{1.120}$$

We now obtain the diagonalized state variable description

$$\dot{\mathbf{x}}(t) = \Lambda \mathbf{x}(t) + \mathbf{b}u(t) \tag{1.121a}$$

$$y(t) = \mathbf{c}^{\mathrm{T}}\mathbf{x}(t) + du(t) \tag{1.121b}$$

where

$$\Lambda = \begin{pmatrix} \lambda_1 & 0 & 0 & \dots & 0 \\ 0 & \lambda_2 & 0 & \dots & 0 \\ 0 & 0 & \lambda_3 & \dots & 0 \\ \vdots & & & \ddots & \vdots \\ 0 & 0 & & \dots & \lambda_n \end{pmatrix}, \quad \mathbf{b} = \begin{pmatrix} 1 \\ 1 \\ 1 \\ \vdots \\ 1 \end{pmatrix}$$

$$\mathbf{c}^{\mathrm{T}} = (\gamma_1 \quad \gamma_2 \quad \dots \quad \gamma_n), \; d = \delta$$

Next we consider a transfer function with poles which are not all distinct. We give an example to explain the procedure to obtain the Jordan block form. Assume that $H(s)$ can be expanded by partial fractions as

$$H(s) = \frac{\beta_1 s^{n-1} + \beta_2 s^{n-2} + \cdots + \beta_n}{(s-\lambda_1)^3 (s-\lambda_4)\dots(s-\lambda_n)} + \delta$$

$$= \frac{\gamma_{11}}{(s-\lambda_1)^3} + \frac{\gamma_{12}}{(s-\lambda_1)^2} + \frac{\gamma_{13}}{(s-\lambda_1)} + \frac{\gamma_4}{(s-\lambda_4)} + \cdots + \frac{\gamma_n}{(s-\lambda_n)} + \delta$$

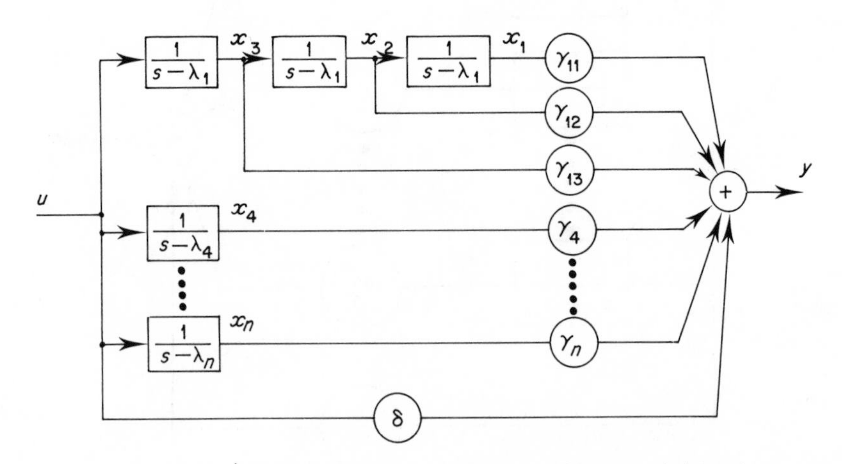

Figure 1.15 Realization by Jordan block form: case (b) repeated eigenvalues

where

$$\gamma_{11} = \lim_{s \to \lambda_1} \{(s - \lambda_1)^3 H(s)\}$$

$$\gamma_{12} = \lim_{s \to \lambda_1} \left\{\frac{d}{ds}(s - \lambda_1)^3 H(s)\right\}$$

$$\gamma_{13} = \lim_{s \to \lambda_1} \left\{\frac{d^2}{ds^2}(s - \lambda_1)^3 H(s)\right\}$$

$$\gamma_i = \lim_{s \to \lambda_i} \{(s - \lambda_i)H(s)\}.$$

We assign the state variables $x_1, x_2, \ldots x_n$, as illustrated in Figure 1.15, then

$$\begin{aligned}
&\dot{x}_3(t) - \lambda_1 x_3(t) = u(t) \\
&\dot{x}_2(t) - \lambda_1 x_2(t) = x_3(t) \\
&\dot{x}_1(t) - \lambda_1 x_1(t) = x_2(t) \\
&\dot{x}_4(t) - \lambda_4 x_4(t) = u(t) \\
&\qquad\vdots \\
&\dot{x}_n(t) - \lambda_n x_n(t) = u(t) \\
&y(t) = \gamma_{11}x_1(t) + \gamma_{12}x_2(t) + \gamma_{13}x_3(t) + \gamma_4 x_4(t) + \cdots + \gamma_n x_n(t) + \delta u(t)
\end{aligned}$$

Rewriting in matrix form, we have the Jordan block canonical form

$$\begin{pmatrix} \dot{x}_1 \\ \dot{x}_2 \\ \dot{x}_3 \\ \dot{x}_4 \\ \vdots \\ \dot{x}_n \end{pmatrix} = \begin{pmatrix} \lambda_1 & 1 & 0 & 0 & \ldots & 0 \\ 0 & \lambda_1 & 1 & 0 & \ldots & 0 \\ 0 & 0 & \lambda_1 & 0 & 0 & 0 \\ 0 & 0 & 0 & \lambda_4 & 0 & \vdots \\ \vdots & & & & \ddots & \\ 0 & & \ldots & & & \lambda_n \end{pmatrix} \begin{pmatrix} x_1 \\ x_2 \\ x_3 \\ x_4 \\ \vdots \\ x_n \end{pmatrix} + \begin{pmatrix} 0 \\ 0 \\ 1 \\ 1 \\ 1 \\ \vdots \\ 1 \end{pmatrix} u \tag{1.122a}$$

$$y = (\gamma_{11} \quad \gamma_{12} \quad \gamma_{13} \quad \gamma_4 \ldots \gamma_n)\mathbf{x} + \delta u \tag{1.122b}$$

It is seen that the matrix A consists of one Jordan block, shown dotted, associated with each eigenvalue. Because λ_1 is of multiplicity three the corresponding Jordan block is three by three. The elements above the diagonal in this block are unity, or in some rather special cases, zero.

Every linear time-invariant system has an equivalent Jordan canonical form. For instance, the non-singular matrix which transforms the companion canonical form (1.115) into the Jordan canonical form (1.121) is given by

$$T = \begin{pmatrix} 1 & 1 & \dots & 1 \\ \lambda_1 & \lambda_2 & \dots & \lambda_n \\ \lambda_1^2 & \lambda_2^2 & \dots & \lambda_n^2 \\ & & \dots & \\ \lambda_1^{n-1} & \lambda_2^{n-1} & \dots & \lambda_n^{n-1} \end{pmatrix}, \ \det T \neq 0 \tag{1.123}$$

where $\lambda_1, \lambda_2, \ldots, \lambda_n$ are the eigenvalues of the companion matrix A in (1.115). This result can be easily proved as follows. With A taken as in (1.115) we have

$$AT = \begin{pmatrix} 0 & 1 & 0 & \dots & 0 \\ 0 & 0 & 1 & \dots & 0 \\ \vdots & & & \ddots & \vdots \\ & & & & 1 \\ -\alpha_0 & -\alpha_1 & & \dots & -\alpha_{n-1} \end{pmatrix} \begin{pmatrix} 1 & 1 & \dots & 1 \\ \lambda_1 & \lambda_2 & \dots & \lambda_n \\ & & \dots & \\ & & \dots & \\ \lambda_1^{n-1} & \lambda_2^{n-1} & \dots & \lambda_n^{n-1} \end{pmatrix}$$

$$= \begin{pmatrix} \lambda_1 & \lambda_2 & \dots & \lambda_n \\ \lambda_1^2 & \lambda_2^2 & \dots & \lambda_n^2 \\ & & \dots & \\ \lambda_1^n & \lambda_2^n & \dots & \lambda_n^n \end{pmatrix}$$

$$= \begin{pmatrix} 1 & 1 & \dots & 1 \\ \lambda_1 & \lambda_2 & \dots & \lambda_n \\ & & \vdots & \\ & & \dots & \\ \lambda_1^{n-1} & \lambda_2^{n-1} & \dots & \lambda_n^{n-1} \end{pmatrix} \begin{pmatrix} \lambda_1 & 0 & \dots & 0 \\ 0 & \lambda_2 & \dots & 0 \\ \vdots & & \ddots & \vdots \\ 0 & & \dots & \lambda_n \end{pmatrix}$$

Thus $AT = T\Lambda$, giving $\Lambda = T^{-1}AT$ and T is therefore the non-singular matrix which transforms the companion form into the equivalent Jordan form.

1.5.4 Tridiagonal Form

It is assumed that the strictly proper part $H_0(s)$ of the transfer function $H(s)$ can be expanded in a continued fraction as

$$H_0(s) = \cfrac{1}{b_1 + a_1 s + \cfrac{1}{b_2 + a_2 s + \cfrac{1}{\ddots \; \cfrac{1}{b_{n-1} + a_{n-1} s + \cfrac{1}{b_n + a_n s}}}}}$$

where $a_i \neq 0$ for $i = 1, \ldots, n$.

If we assign the state variables from

$$\frac{X_n(s)}{X_{n-1}(s)} = \frac{1}{b_n + a_n s}$$

$$\frac{X_{n-1}(s)}{X_{n-2}(s)} = \frac{1}{b_{n-1} + a_{n-1}s + \dfrac{X_n(s)}{X_{n-1}(s)}}$$

$$\cdots$$

$$\frac{X_2(s)}{X_1(s)} = \frac{1}{b_2 + a_2 s + \dfrac{X_3(s)}{X_2(s)}}$$

$$\frac{X_1(s)}{U(s)} = \frac{1}{b_1 + a_1 s + \dfrac{X_2(s)}{X_1(s)}}$$

$$Y(s) = X_1(s)$$

we then have the following state equations

$$\dot{x}_n = \frac{1}{a_n} x_{n-1} - \frac{b_n}{a_n} x_n$$

$$\dot{x}_{n-1} = \frac{1}{a_{n-1}} x_{n-2} - \frac{b_{n-1}}{a_{n-1}} x_{n-1} - \frac{1}{a_{n-1}} x_n$$

$$\cdots$$

$$\dot{x}_2 = \frac{1}{a_2} x_1 - \frac{b_2}{a_2} x_2 - \frac{1}{a_2} x_3$$

$$\dot{x}_1 = -\frac{b_1}{a_1} x_1 - \frac{1}{a_1} x_2 + \frac{1}{a_1} u.$$

They can be rewritten in matrix form as

$$\begin{pmatrix} \dot{x}_1 \\ \dot{x}_2 \\ \vdots \\ \\ \dot{x}_n \end{pmatrix} = \begin{pmatrix} -\frac{b_1}{a_1} & -\frac{1}{a_1} & 0 & \cdots & 0 \\ \frac{1}{a_2} & -\frac{b_2}{a_2} & -\frac{1}{a_2} & 0 & \vdots \\ 0 & & \cdots & & \\ \vdots & & \frac{1}{a_{n-1}} & -\frac{b_{n-1}}{a_{n-1}} & -\frac{1}{a_{n-1}} \\ 0 & \cdots & 0 & \frac{1}{a_n} & -\frac{b_n}{a_n} \end{pmatrix} \begin{pmatrix} x_1 \\ x_2 \\ \vdots \\ \\ x_n \end{pmatrix} + \begin{pmatrix} \frac{1}{a_1} \\ 0 \\ \vdots \\ \\ 0 \end{pmatrix} u$$

$$y = (1 \quad 0 \quad \dots \quad 0)\mathbf{x}$$

PROBLEMS

P1.1 Find, by use of the Faddeev algorithm, the transfer function of the linear system $(A, \mathbf{b}, \mathbf{c}^{\mathrm{T}})$ given by

$$A = \begin{pmatrix} -2 & 0 & 1 \\ 0 & -2 & 1 \\ 0 & -3 & -2 \end{pmatrix}, \mathbf{b} = \begin{pmatrix} 1 \\ 2 \\ 2 \end{pmatrix}, \mathbf{c}^{\mathrm{T}} = (-1 \quad 1 \quad 0)$$

P1.2 Find the state variable description of the network given in Figure P1.2, where the input $u(t)$ is the source voltage and the output voltage $y(t)$ is the voltage across the resistor R_3.

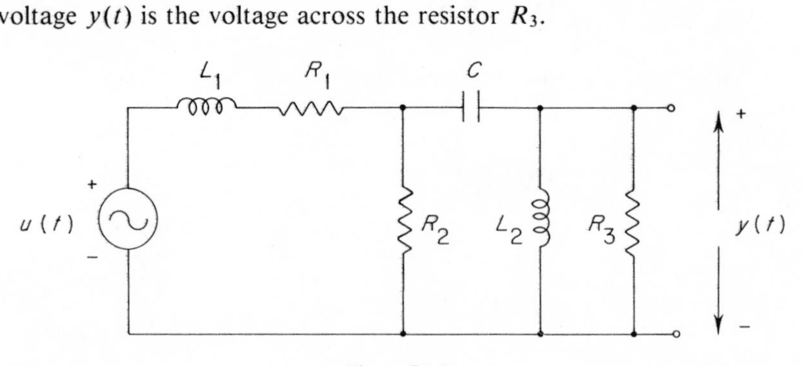

Figure P1.2

P1.3 Construct the state variable description of the tank system shown in Figure P1.3, where input $u(t)$ is the incoming flow rate, the output $y(t)$ the outgoing flow rate, S_1, S_2 and S_3 the cross-sectional areas of the tanks and

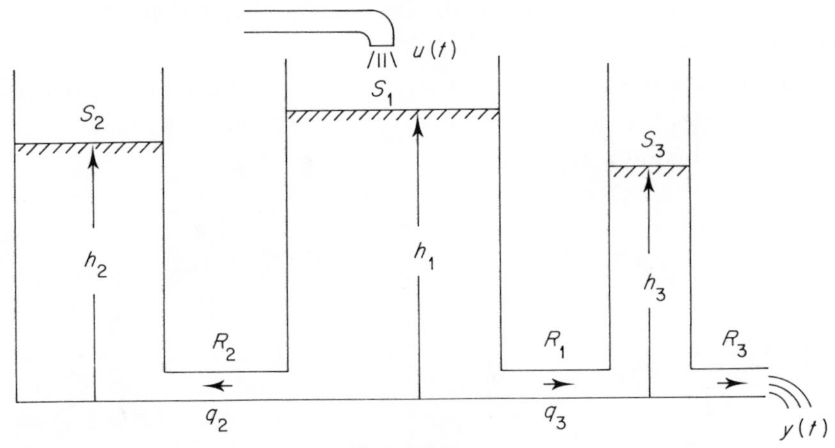

Figure P1.3

h_1, h_2 and h_3 the levels. It is assumed that the each flow rate q_2, q_3 and y through the pipes is proportional to the difference in the levels, that is, $q_2 = (h_1 - h_2)/R_2$, where R_2 is the flow resistance.

P1.4 Write node voltage equations for the circuit shown in Figure P1.4 and show that it is the force current analogue of Figure 1.7(a).

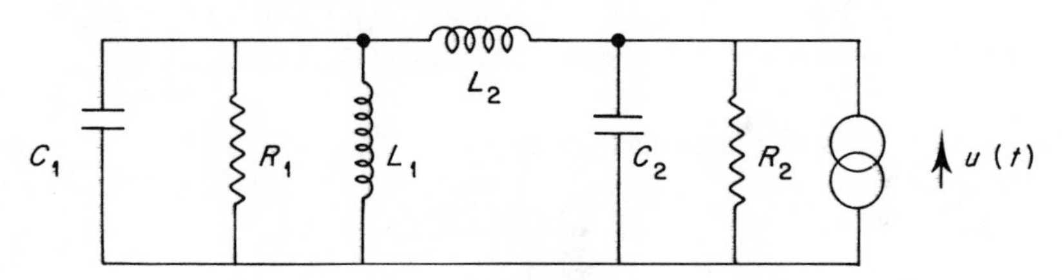

Figure P1.4

P1.5 Obtain state variable equations for the network shown in Figure P1.5 using the voltages across the capacitors and the current through

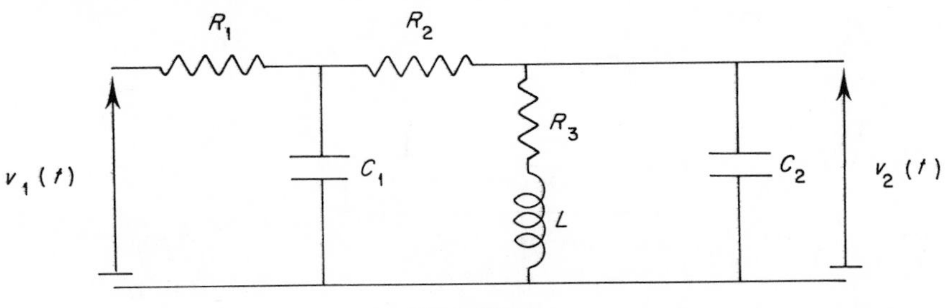

Figure P1.5

the inductance as the state variables. Determine the A, B, C and D matrices if $v_1(t)$ is the input and $v_2(t)$ the output.

P1.6 Write the state equations for the network of Figure P1.6 using the state variables i_1, i_2 and v, e_1 and e_2 are the network inputs and i_1 and i_2 are the required outputs.

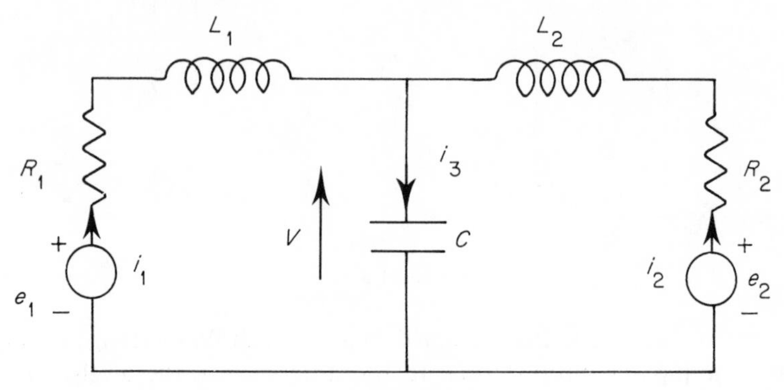

Figure P1.6

P1.7 Obtain the differential equations for the system shown in Figure P1.7. Draw a block diagram for this plant and find the transfer function $X_2(s)/F(s)$. Write the state-variable representation for the plant, using the variables $z_1 = x_1$, $z_2 = \dot{x}_1$, $z_3 = x_2$, and $z_4 = \dot{x}_2$.

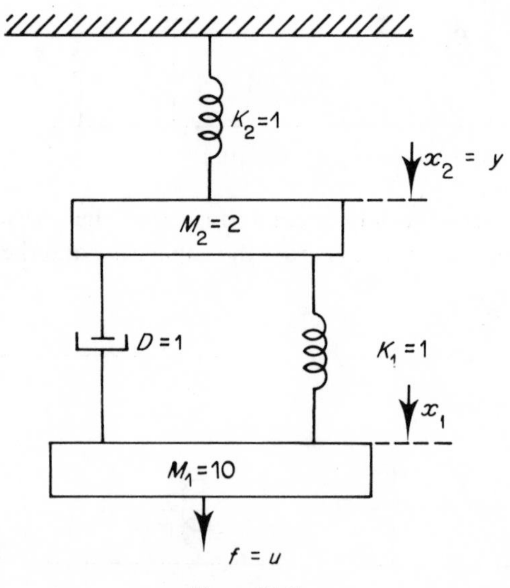

Figure P1.7

P1.8 Draw the electric circuit which is the force-current analogue of the mass, spring, damper system of Figure P1.7.

P1.9 Show that the circuit shown in Figure P1.9 has a transfer function from v_3 to v_2 of

$$\frac{V_2(s)}{V_3(s)} = \frac{1}{s^3LRC^2 + 2s^2LC + s(2CR + L/R) + 2}$$

and it can be represented in state variable form with

$$A = \begin{pmatrix} 0 & 1 & 0 \\ 0 & 0 & 1 \\ -2a^2b & -a(a+2b) & -2a \end{pmatrix}, B = \begin{pmatrix} 0 \\ 0 \\ a^2b \end{pmatrix}, \text{ and } C = (1 \quad 0 \quad 0)$$

where $a = 1/RC$ and $b = R/L$.

Choosing the state variables $x_1 = v_1$, $x_2 = v_2$ and $x_3 = Li_3$ show that the new A matrix A' is given by

$$A' = \begin{pmatrix} -a & -a & -ab \\ 0 & -a & ab \\ 1 & -1 & 0 \end{pmatrix}$$

Determine the transformation matrix T relating the state variables for the two choices of the A matrix.

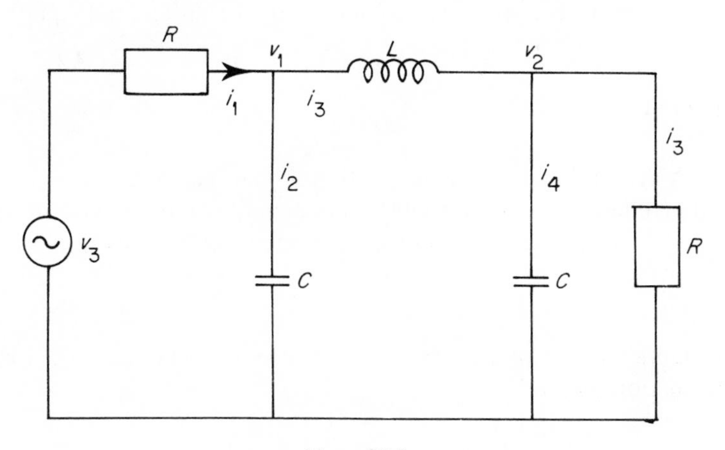

Figure P1.9

P1.10 Obtain the state space representation of the network shown in Figure P1.10 with the currents through the inductors as state variables, u_1 and u_2 the input voltages and y_1 and y_2 as the output voltages.

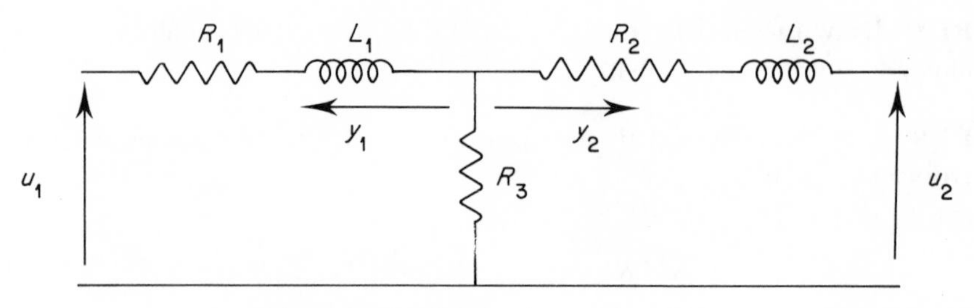

Figure P1.10 $R_i = 1$ and $L_i = 2$ for all i

P1.11 Write in controllable canonical form the state equation for the differential equation

$$\ddot{y} + 3\dot{y} + 2y = u(t).$$

and draw the corresponding signal flow graph. If a state variable transformation $\mathbf{x} = T\bar{\mathbf{x}}$ is used, where $\mathbf{x}$ and $\bar{\mathbf{x}}$ are respectively the old and the new state variables, determine the new state equations for the differential equation if

$$T = \begin{pmatrix} 1 & 1 \\ -1 & -2 \end{pmatrix}.$$

Draw a signal flow graph for this new representation.

P1.12 Obtain a state space representation of the differential equation

$$\dddot{x} + 3\ddot{x} + 3\dot{x} + x = \dot{u} + u$$

with the output $y = x$.

P1.13 A system has an A matrix with a modal matrix U (i.e. the matrix which diagonalizes A). Show that the modal matrix V of the system with A matrix A^{T} is, with a suitable choice of normalizing constants, given by

$$V^{\mathrm{T}} = U^{-1}$$

P1.14 Obtain the response $y(t)$ of the following system using the Laplace transformation method

$$\dot{\mathbf{x}} = \begin{pmatrix} -3 & -2 & -1 \\ 1 & 0 & -1 \\ 0 & 0 & -4 \end{pmatrix} \mathbf{x} + \begin{pmatrix} 1 \\ 0 \\ 0 \end{pmatrix} u$$

$$y = x_1 + 2x_2 + 3x_3.$$

u is a unit step function and $\mathbf{x}(0) = \mathbf{0}$.

P1.15 A dynamic system is described by the state space equation

$$\dot{\mathbf{x}} = A\mathbf{x} + \mathbf{b}u \qquad \mathbf{x}(0) = \mathbf{0}$$
$$y = x_1$$

where

$$A = \begin{pmatrix} 8 & -8 & -2 \\ 4 & -3 & -2 \\ 3 & -4 & 1 \end{pmatrix} \quad \text{and} \quad \mathbf{b} = \begin{pmatrix} 0 \\ 0 \\ 1 \end{pmatrix}$$

(a) Find a non-singular transformation T such that $T^{-1}AT$ is diagonal.
(b) From (a), find e^{At}.
(c) Find e^{At} using the Sylvester expansion.
(d) Find y if u is a unit impulse function.

P1.16 A system is governed by the state equation

$$\dot{\mathbf{x}}(t) = \begin{pmatrix} -3 & 1 \\ 1 & -3 \end{pmatrix}\mathbf{x}(t) + \begin{pmatrix} 1 & 1 \\ 1 & 1 \end{pmatrix}\mathbf{u}(t),$$

where $\mathbf{x}(t)$ and $\mathbf{u}(t)$ are respectively the state and input vectors of the system. Determine the transition matrix of this system and hence obtain an explicit expression for $\mathbf{x}(t)$ if

$$\mathbf{x}(0) = \begin{pmatrix} 1 \\ 2 \end{pmatrix} \text{ and } \mathbf{u}(t) = \begin{pmatrix} 4 \\ 3 \end{pmatrix} (t > 0)$$

P1.17 For the system shown in Figure P1.17 write a state space description with x_1 and x_2 the state variables. If $u(t) = 0$ for all t and $\mathbf{x}(2) = (1 \; 0)^{\mathrm{T}}$, find $\mathbf{x}(0)$ and $\mathbf{x}(3)$

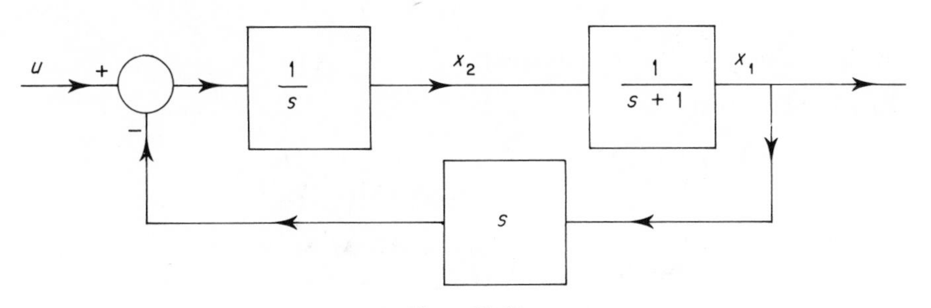

Figure P1.17

P1.18 Find the transition matrix of A if

$$A = \begin{pmatrix} 0 & 1 & 0 \\ 0 & 0 & 1 \\ -2 & -4 & -3 \end{pmatrix}.$$

P1.19 The state space equation of a system is

$$\dot{\mathbf{x}} = \begin{pmatrix} 0 & 1 & 0 \\ 3 & 0 & 2 \\ -12 & -7 & -6 \end{pmatrix} \mathbf{x} + \begin{pmatrix} -1 \\ 2 \\ 3 \end{pmatrix} u$$

where $\mathbf{x} = (x_1, x_2, x_3)^\mathrm{T}$. Obtain the scalar differential equation satisfied by x_1.

P1.20 Represent the following plants by means of state variable descriptions with A in the controllable canonical form.

(a) $G_\mathrm{p}(s) = \dfrac{1}{s^2 + 2s + 3}$

(b) $G_\mathrm{p}(s) = \dfrac{s + 2}{s^3 + 3s^2 + 2s + 10}$

(c) $G_\mathrm{p}(s) = \dfrac{10(s + 1)(s + 2)}{s(s + 3)(s + 4)}$

P1.21 The transfer function $G(s)$ is given by

$$G(s) = \frac{s^2 + 7s + 10}{s^3 + 8s^2 + 19s + 122}$$

Find (a) the controllable canonical form, (b) the observable canonical form, (c) the Jordan canonical form and (d) the tridiagonal form. Draw the corresponding block diagrams.

P1.22 Verify that

$$\mathrm{e}^{(A+B)t} = \mathrm{e}^{At}\mathrm{e}^{Bt}$$

holds true only if A and B commute, i.e. $AB = BA$.

P1.23 Show that

$$\begin{pmatrix} \mathbf{A}_{11} & \mathbf{O} \\ \mathbf{A}_{21} & \mathbf{A}_{22} \end{pmatrix}^{-1} = \begin{pmatrix} \mathbf{A}_{11}^{-1} & \mathbf{O} \\ -\mathbf{A}_{22}^{-1}\mathbf{A}_{21}\mathbf{A}_{11}^{-1} & \mathbf{A}_{22}^{-1} \end{pmatrix}$$

$$\begin{pmatrix} \mathbf{A}_{11} & \mathbf{A}_{12} \\ \mathbf{O} & \mathbf{A}_{22} \end{pmatrix}^{-1} = \begin{pmatrix} \mathbf{A}_{11}^{-1} & -\mathbf{A}_{11}^{-1}\mathbf{A}_{12}\mathbf{A}_{22}^{-1} \\ \mathbf{O} & \mathbf{A}_{22}^{-1} \end{pmatrix}$$

where $\mathbf{A}_{11}$ and $\mathbf{A}_{22}$ are assumed to be non-singular.

P1.24 Show that

$$\begin{pmatrix} \mathbf{A}_{11} & \mathbf{A}_{12} \\ \mathbf{A}_{21} & \mathbf{A}_{22} \end{pmatrix}^{-1} = \begin{pmatrix} \mathbf{A}_{11}^{-1} + \mathbf{A}_{12}^{-1}\mathbf{A}_{12}\mathbf{D}^{-1}\mathbf{A}_{21}\mathbf{A}_{11}^{-1} & -\mathbf{A}_{11}^{-1}\mathbf{A}_{12}\mathbf{D}^{-1} \\ -\mathbf{D}^{-1}\mathbf{A}_{21}\mathbf{A}_{11}^{-1} & \mathbf{D}^{-1} \end{pmatrix}$$

where $\mathbf{A}_{11}$ and $\mathbf{D} \equiv \mathbf{A}_{22} - \mathbf{A}_{21}\mathbf{A}_{11}^{-1}\mathbf{A}_{12}$ are assumed to be non-singular.

2
STRUCTURE OF LINEAR SYSTEMS

2.1 OBSERVABILITY AND CONTROLLABILITY OF TIME-INVARIANT SYSTEMS

Let us consider the time-invariant system represented by

$$\dot{\mathbf{x}}(t) = A\mathbf{x}(t) + B\mathbf{u}(t) \tag{2.1a}$$

$$\mathbf{y}(t) = C\mathbf{x}(t) + D\mathbf{u}(t) \tag{2.1b}$$

where $\mathbf{x}$, $\mathbf{u}$, $\mathbf{y}$ are an n-vector, an m-vector and a p-vector. The objective of the control is to transfer the state of the system to a desirable state from the initial state using the input $\mathbf{u}$. However, the existence of such an input should be assured; this is the controllability condition. On the other hand, it is sometimes necessary to know all state variables from measurement of the output $\mathbf{y}(t)$ whose dimension is less than that of the state. The observability condition assures the construction of the state from the output. These properties are intrinsic for systems and play important roles in linear system theory.

2.1.1 The Condition for Controllability

Definition 2.1 (Controllability). For the linear system given by (2.1) if there exists an input $\mathbf{u}_{[0,t_1]}$ which transfers the initial state $\mathbf{x}(0) = \mathbf{x}_0$ to the zero state $\mathbf{x}(t_1) = \mathbf{0}$ in a finite time t_1, the state $\mathbf{x}_0$ is said to be controllable. If all initial states are controllable the system is said to be completely controllable.

From (1.73), the solution of (2.1a) is

$$\mathbf{x}(t) = \mathrm{e}^{At}\mathbf{x}_0 + \int_0^t \mathrm{e}^{A(t-\tau)}B\mathbf{u}(\tau)\,\mathrm{d}\tau \tag{2.2}$$

If the system is controllable, i.e., there exists an input to make $\mathbf{x}(t_1) = \mathbf{x}_1 = 0$ at finite time $t = t_1$, then after premultiplying by e^{-At_1}, (2.2) yields

$$-\mathbf{x}_0 = \int_0^{t_1} \mathrm{e}^{-A\tau}B\mathbf{u}(\tau)\,\mathrm{d}\tau \tag{2.3}$$

Therefore any controllable state satisfies (2.3), and for a completely controllable system every state $\mathbf{x}_0 \in R^n$ satisfies (2.3) with t_1 (>0) and $\mathbf{u}_{[0,t_1]}$.†

From (2.3), it is found that complete controllability of a system depends on A and B, and is independent of the output matrix C.

The theorem to check the controllability of the system is given as follows.

Theorem 2.1

The necessary and sufficient condition for the system (2.1) to be completely controllable is given by one of the following conditions:

(i) $W(0, t_1) = \int_0^{t_1} \mathrm{e}^{-At} BB^{\mathrm{T}} \mathrm{e}^{-A^{\mathrm{T}}t}\, \mathrm{d}t$ is non-singular,
(ii) The controllability matrix $\mathscr{C} \triangleq [B, AB, A^2B, \ldots, A^{n-1}B]$, $(n \times nm)$ satisfies rank $\mathscr{C} = n$.‡

Since condition (ii) can be computed without integration it allows the controllability of a system to be easily checked. As seen from Theorem 2.1, the complete controllability of the system comes from the properties of A and B. So one simply states that '(A, B) is controllable'. Before the proof is done some examples are given.

Example 2.1 The system represented by

$$\dot{\mathbf{x}} = \begin{pmatrix} 0 & 0 & -1 \\ 1 & 0 & -3 \\ 0 & 1 & -3 \end{pmatrix} \mathbf{x} + \begin{pmatrix} 1 \\ 1 \\ 0 \end{pmatrix} u$$

is considered. The rank of the controllability matrix of the given system is 2 as shown below. So the system is not completely controllable.

$$\text{rank } \mathscr{C} = \text{rank}[\mathbf{b}, A\mathbf{b}, A^2\mathbf{b}] = \text{rank} \begin{pmatrix} 1 & 0 & -1 \\ 1 & 1 & -3 \\ 0 & 1 & -2 \end{pmatrix} = 2 \neq 3$$

Example 2.2 The state equation of the circuit shown in Figure 2.1 can be derived using the method described in Section 1.2.4. Let the state variables be the current of

†R denotes the set of real numbers $\{x \mid -\infty < x < \infty\}$ and R^n gives an n-dimensional real vector space. $\mathbf{x} \in R^n$ denotes that $\mathbf{x}$ is an element of R^n, i.e., $\mathbf{x}$ is an n-dimensional real vector.
‡Rank $\mathscr{C}$ denotes the rank of the matrix $\mathscr{C}$, which is equal to the number of linearly independent column (or row) vectors in $\mathscr{C}$.

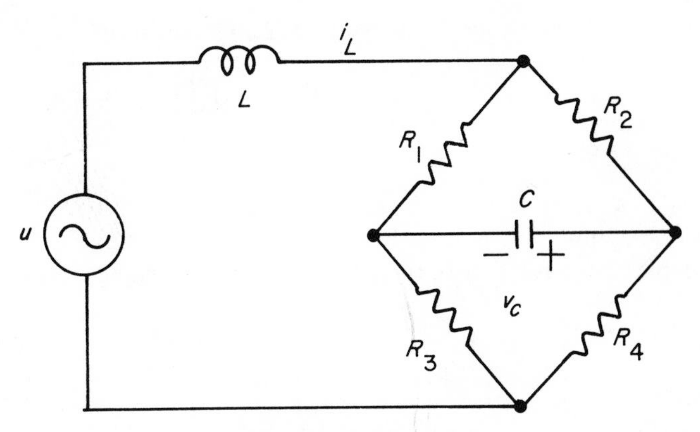

Figure 2.1 Bridge circuit

the inductance, i_L, and the voltage of the capacitance v_c, and let the output be i_L, then

$$\begin{pmatrix} \dot{i}_L \\ \dot{v}_C \end{pmatrix} = \begin{pmatrix} -\dfrac{1}{L}\left[\dfrac{R_1R_2}{R_1+R_2}+\dfrac{R_3R_4}{R_3+R_4}\right], & -\dfrac{1}{L}\left[\dfrac{R_1}{R_1+R_2}-\dfrac{R_3}{R_3+R_4}\right] \\ -\dfrac{1}{C}\left[\dfrac{R_2}{R_1+R_2}-\dfrac{R_4}{R_3+R_4}\right], & -\dfrac{1}{C}\left[\dfrac{1}{R_1+R_2}+\dfrac{1}{R_3+R_4}\right] \end{pmatrix}\begin{pmatrix} i_L \\ v_C \end{pmatrix} + \begin{pmatrix} \dfrac{1}{L} \\ 0 \end{pmatrix} u$$

$$y = (1 \quad 0)\begin{pmatrix} i_L \\ v_C \end{pmatrix}.$$

The controllability matrix of the system is

$$\mathscr{C} = [\mathbf{b}, A\mathbf{b}] = \begin{pmatrix} \dfrac{1}{L} & -\dfrac{1}{L^2}\left[\dfrac{R_1R_2}{R_1+R_2}+\dfrac{R_3R_4}{R_3+R_4}\right] \\ 0 & \dfrac{1}{LC}\left[\dfrac{R_4}{R_3+R_4}-\dfrac{R_2}{R_1+R_2}\right] \end{pmatrix}.$$

Thus we see that under the condition that $R_4/(R_3+R_4) = R_2/(R_1+R_2)$, that is, $R_1R_4 = R_2R_3$, rank $\mathscr{C} = 1$ and the system becomes 'uncontrollable'. This condition is the one required to balance the resistance bridge, and in this case, the voltage of the capacitance v_c cannot be varied by any external input u.

Proof of Theorem 2.1 (due to Brockett) Condition (i): *Sufficiency*. If $W(0, t_1)$ given in (i) is non-singular, the following input can be applied to the system

$$\mathbf{u}(t) = -B^{\mathrm{T}}\mathrm{e}^{-A^{\mathrm{T}}t}W^{-1}(0, t_1)\mathbf{x}_0 \tag{2.4}$$

For the input (2.4), the state of the system (2.2) is given by

$$\begin{aligned}\mathbf{x}(t_1) &= e^{At_1}\mathbf{x}_0 - e^{At_1}\left\{\int_0^{t_1} e^{-A\tau}BB^{\mathrm{T}}e^{-A^{\mathrm{T}}\tau}\,d\tau\right\}W^{-1}(0, t_1)\mathbf{x}_0 \\ &= \mathbf{0}\end{aligned} \tag{2.5}$$

for any initial state $\mathbf{x}_0$. Therefore the system (A, B) is controllable.

Necessity. Assume that though $W(0, t_1)$ is singular for any t_1, the system is controllable. Then for $t_1 > 0$, there exists a non-zero n-vector $\boldsymbol{\alpha}$ such that

$$\boldsymbol{\alpha}^{\mathrm{T}} W(0, t_1)\boldsymbol{\alpha} = \int_0^{t_1} \boldsymbol{\alpha}^{\mathrm{T}} e^{-A\tau}BB^{\mathrm{T}}e^{-A^{\mathrm{T}}\tau}\boldsymbol{\alpha}\,d\tau = 0 \tag{2.6}$$

which yields for any $t \geqslant 0$

$$\boldsymbol{\alpha}^{\mathrm{T}}e^{-At}B = \mathbf{0}^{\mathrm{T}}, \qquad t \geqslant 0 \tag{2.7}$$

From the assumption of controllability, there exists an input $\mathbf{u}$ satisfying (2.3), therefore from (2.3) and (2.7)

$$-\boldsymbol{\alpha}^{\mathrm{T}}\mathbf{x}_0 = \boldsymbol{\alpha}^{\mathrm{T}} \int_0^{t_1} e^{-A\tau}B\mathbf{u}(\tau)\,d\tau = 0 \tag{2.8}$$

holds for any initial state $\mathbf{x}_0$. By choosing $\mathbf{x}_0 = \boldsymbol{\alpha}$, (2.8) gives $\boldsymbol{\alpha} = 0$ which contradicts the non-zero property of $\boldsymbol{\alpha}$. Therefore the non-singularity of $W(0, t_1)$ is proved.

Condition (ii): *Sufficiency.* It is first assumed that though rank $\mathscr{C} = n$ the system is not controllable, and by showing that this is a contradiction the controllability of the system is proved. By the above assumption that the system is not controllable and rank $\mathscr{C} = n$, $W(0, t_1)$ is singular. Therefore (2.7), that is

$$\boldsymbol{\alpha}^{\mathrm{T}}e^{-At}B = \mathbf{0}^{\mathrm{T}}, \qquad t \geqslant 0,\ \boldsymbol{\alpha} \neq 0 \tag{2.9}$$

holds. Derivatives of the above equation at $t = 0$ yield

$$\boldsymbol{\alpha}^{\mathrm{T}}A^kB = \mathbf{0}^{\mathrm{T}}, \qquad k = 0, 1, \ldots, n-1 \tag{2.10}$$

which is equivalent to

$$\boldsymbol{\alpha}^{\mathrm{T}}[B, AB, \ldots, A^{n-1}B] = \boldsymbol{\alpha}^{\mathrm{T}}\mathscr{C} = \mathbf{0}^{\mathrm{T}} \tag{2.11}$$

This contradicts the assumption that rank $\mathscr{C} = n$ so the system is completely controllable.

Necessity. It is assumed that the system is completely controllable, but rank $\mathscr{C} < n$. From the assumption there exists a non-zero $\boldsymbol{\alpha}$ satisfying (2.11), or equivalently (2.10). Also from the Cayley-Hamilton theorem, A^{n+i} can be expressed as a linear combination of $I, A, \ldots, A^{n-1}$ so that e^{-At} may be expressed as a linear combination of $I, A, \ldots, A^{n-1}$, which yields

$$\boldsymbol{\alpha}^{\mathrm{T}}e^{-At}B = \mathbf{0}^{\mathrm{T}}, \qquad t \geqslant 0,\ \boldsymbol{\alpha} \neq 0 \tag{2.12}$$

and therefore

$$\mathbf{0} = \int_0^{t_1} \boldsymbol{\alpha}^{\mathrm{T}} \mathrm{e}^{-A\tau} BB^{\mathrm{T}} \mathrm{e}^{-A^{\mathrm{T}}\tau} \boldsymbol{\alpha} \, \mathrm{d}\tau = \boldsymbol{\alpha}^{\mathrm{T}} W(0, t_1) \boldsymbol{\alpha}. \tag{2.13}$$

Since the system is completely controllable, $W(0, t_1)$ should be non-singular from (i). There $\boldsymbol{\alpha}$ of (2.13) should be zero, which contradicts the assumption that $\boldsymbol{\alpha}$ is non-zero. So rank $\mathscr{C} = n$.

2.1.2 The Condition for Observability

Observability is concerned with the output equation of the system.

Definition 2.2

Observability. When using the output of the system (2.1) measured from time 0 to time t_1, if the initial state $\mathbf{x}(0) = \mathbf{x}_0$ is uniquely determined, $\mathbf{x}_0$ is said to be observable, when the input is assumed to be completely known. When all states are observable, the system is said to be completely observable, or (A, C) is observable.

The output of the system (2.1) is given by

$$\mathbf{y}(t) = C\mathrm{e}^{At}\mathbf{x}_0 + C \int_0^t \mathrm{e}^{A(t-\tau)} B\mathbf{u}(\tau) \, \mathrm{d}\tau + D\mathbf{u}(t) \tag{2.14}$$

The output and the input can be measured and used, so the following signal $\boldsymbol{\eta}$ can be obtained from $\mathbf{u}$ and $\mathbf{y}$.

$$\boldsymbol{\eta}(t) \triangleq \mathbf{y}(t) - C \int_0^t \mathrm{e}^{A(t-\tau)} B\mathbf{u}(\tau) \, \mathrm{d}\tau - D\mathbf{u}(t) = C\mathrm{e}^{At}\mathbf{x}_0 \tag{2.15}$$

where C is a $p \times n$ matrix.

Since p is usually less than n, $\mathbf{x}_0$ cannot be determined uniquely from $\boldsymbol{\eta}(t)$ at a specific time t. But when the signal $\boldsymbol{\eta}(t)$ is available over a time interval from 0 to t_1 and the system is completely observable, the initial state $\mathbf{x}_0$ can be uniquely determined.

Premultiplying (2.15) by $\mathrm{e}^{At}C^{\mathrm{T}}$ and integrating from 0 to t_1, gives

$$\left\{ \int_0^{t_1} \mathrm{e}^{A^{\mathrm{T}}t} C^{\mathrm{T}} C \mathrm{e}^{At} \, \mathrm{d}t \right\} \mathbf{x}_0 = \int_0^{t_1} \mathrm{e}^{A^{\mathrm{T}}t} C^{\mathrm{T}} \boldsymbol{\eta}(t) \, \mathrm{d}t. \tag{2.16}$$

If the $n \times n$ matrix defined by

$$M(0, t_1) \triangleq \int_0^{t_1} \mathrm{e}^{A^{\mathrm{T}}t} C^{\mathrm{T}} C \mathrm{e}^{At} \, \mathrm{d}t \tag{2.17}$$

is non-singular, $\mathbf{x}_0$ is determined uniquely from (2.16) as

$$\mathbf{x}_0 = M^{-1}(0, t_1) \int_0^{t_1} e^{A^T t} C^T \boldsymbol{\eta}(t) \, dt \tag{2.18}$$

Therefore the non-singularity of $M(0, t_1)$ for $t_1 \geqslant 0$ is a sufficient condition for the system to be completely observable.

On the other hand, if the system is completely observable, $M(0, t_1)$ is non-singular. This will be proved by contradiction as follows. Assume that the system is completely observable but $M(0, t_1)$ is singular, there exists a non-zero vector $\boldsymbol{\alpha}$ such that

$$\boldsymbol{\alpha}^T M(0, t_1) \boldsymbol{\alpha} = 0. \tag{2.19}$$

By substituting (2.17) into (2.19)

$$0 = \boldsymbol{\alpha}^T M(0, t_1) \boldsymbol{\alpha} = \boldsymbol{\alpha}^T \int_0^{t_1} e^{A^T t} C^T C e^{At} \, dt \boldsymbol{\alpha} \tag{2.20}$$

which gives for any time $t \geqslant 0$

$$C e^{At} \boldsymbol{\alpha} = 0, \qquad t \geqslant 0 \tag{2.21}$$

This means that the output is always zero for the non-zero initial condition $\mathbf{x}(0) = \boldsymbol{\alpha} \neq 0$, and the state $\boldsymbol{\alpha}$ cannot be distinguished from the zero state from the measured output, which contradicts the observability of the system. Therefore $M(0, t_1)$ should be non-singular for the system to be completely observable.

From the above results, it is clear that observability is dependent only on the properties of the matrices A and C, and the following theorem concerning observability is given.

Theorem 2.2

A necessary and sufficient condition for the system (2.1) to be completely observable is one of the following equivalent conditions.

(i) $M(0, t_1) = \int_0^{t_1} e^{A^T t} C^T C e^{At} \, dt$ is non-singular.
(ii) The observability matrix defined as the $n \times np$ matrix $\mathcal{O} \triangleq [C^T, A^T C^T, \ldots, (A^T)^{n-1} C^T]$ has rank n.

(i) has been proved before the theorem and (ii) can be proved from (i) in a similar manner to Theorem 2.1.

Example 2.3 The observability matrix of the bridge circuit given in Example 2.2 is given by

$$\mathcal{O} = [\mathbf{c}^{\mathrm{T}}, A^{\mathrm{T}}\mathbf{c}^{\mathrm{T}}] = \begin{pmatrix} 1 & -\dfrac{1}{L}\left[\dfrac{R_1R_2}{R_1+R_2} + \dfrac{R_3R_4}{R_3+R_4}\right] \\ 0 & -\dfrac{1}{L}\left[\dfrac{R_1}{R_1+R_2} - \dfrac{R_3}{R_3+R_4}\right] \end{pmatrix}$$

For the balance condition of $R_1R_4 = R_2R_3$, rank $\mathcal{O} = 1$ and the system is not observable

2.1.3 Duality

The controllability matrix $\mathscr{C}$ and the observability matrix $\mathcal{O}$ have similar structure in that by replacing B and A by C^{T} and A^{T}, $\mathscr{C}$ becomes $\mathcal{O}$.

Let us consider the system

$$\dot{\mathbf{x}}^*(t) = -A^{\mathrm{T}}\mathbf{x}^*(t) - C^{\mathrm{T}}\mathbf{u}^*(t) \tag{2.22a}$$

$$\mathbf{y}^*(t) = B^{\mathrm{T}}\mathbf{x}^*(t) + D^{\mathrm{T}}\mathbf{u}^*(t). \tag{2.22b}$$

By changing $t \to -t$, $A \to A^{\mathrm{T}}$, $B \to C^{\mathrm{T}}$, $C \to B^{\mathrm{T}}$, $D \to D^{\mathrm{T}}$ in the system (2.1), the system (2.22) is obtained, and this system is said to be the dual system to the system (2.1).

From Theorem 2.1 and Theorem 2.2, if the system (2.1) is controllable (observable), its dual system is observable (controllable). This shows that controllability and observability have a dual relationship, and this relationship will again appear between the optimal regulator and the optimal filter (see Section 6.2.1).

2.1.4 Output Controllability

Controllability is concerned with the transfer of the state to the zero state and is irrelevant with respect to the output. In this section we consider the problem of making the output $\mathbf{y}(t)$ zero in a finite time.

Definition 2.3

Output controllability. If there exists an input which makes the output of the system, $\mathbf{y}(t)$, zero in a finite time t, when the output at time zero is arbitrary, then the system is said to be completely output controllable.

Concerning output controllability we give the following theorem similar to Theorem 2.1.

Theorem 2.3

A necessary and sufficient condition for the system (2.1) to be completely output controllable is one of the following conditions.

(i) $W'(0, t_1) = \int_0^{t_1} Ce^{-At}BB^{\mathrm{T}}e^{-A^{\mathrm{T}}t}C^{\mathrm{T}}\,\mathrm{d}t$ is non-singular.
(ii) The output controllability matrix a $p \times nm$ matrix defined by $\mathscr{C}' \triangleq [CB, CAB, CA^2B, \ldots, CA^{n-1}B]$ has rank p.

2.1.5 Equivalent Systems

By the non-singular transformation

$$\mathbf{x} = T\bar{\mathbf{x}}$$

the system (2.1) is transformed into an equivalent system

$$\dot{\bar{\mathbf{x}}}(t) = \bar{A}\bar{\mathbf{x}}(t) + \bar{B}\mathbf{u}(t) \tag{2.23a}$$

$$\mathbf{y}(t) = \bar{C}\bar{\mathbf{x}}(t) + \bar{D}\mathbf{u}(t) \tag{2.23b}$$

as described in Section 1.4.2, where

$$\bar{A} = T^{-1}AT, \quad \bar{B} = T^{-1}B, \quad \bar{C} = CT, \quad \bar{D} = D. \tag{2.24}$$

The controllability matrix of (2.23) is

$$\begin{aligned}\bar{\mathscr{C}} &= [\bar{B}, \bar{A}\bar{B}, \ldots, \bar{A}^{n-1}\bar{B}] = [T^{-1}B, T^{-1}AB, \ldots, T^{-1}A^{n-1}B] \\ &= T^{-1}(\mathscr{C})\end{aligned} \tag{2.25}$$

Since T^{-1} is non-singular

$$\operatorname{rank} \bar{\mathscr{C}} = \operatorname{rank} \mathscr{C}.$$

A similar relationship can be shown for the observability matrices so that the following theorem exists.

Theorem 2.4

Complete controllability and observability are preserved for equivalent systems.

2.2 STATE SPACE STRUCTURE OF TIME-INVARIANT SYSTEMS

The conditions for a system to be completely controllable or completely observable have been given in Theorem 2.1 or Theorem 2.2, and the concepts have been demonstrated by Examples 2.1, 2.2 and 2.3. This section describes the state space structure from the viewpoint of controllability and observability.

2.2.1 Controllable Subspace

The system (2.1) is assumed not to satisfy the complete controllability condition of Theorem 2.1, then although all states may not be controllable there may exist some controllable states. Since a controllable state is represented by (2.3), the set of all controllable states $\mathscr{X}_c$, named the controllable subspace, is given by

$$\mathscr{X}_c = \left\{ \mathbf{x}_0 \mid \mathbf{x}_0 = -\int_0^t e^{-A\tau} B\mathbf{u}(\tau)\, d\tau,\ ^\forall\mathbf{u} \right\} \tag{2.26}$$ †

Since for $\mathbf{x}_1,\ \mathbf{x}_2 \in \mathscr{X}_c$, $\mathbf{x}_1 + \mathbf{x}_2 \in \mathscr{X}_c$, and for $\mathbf{x}_1 \in \mathscr{X}_c$ and a real number α, $\alpha\mathbf{x}_1 \in \mathscr{X}_c$, then $\mathscr{X}_c$ is a linear space. The controllability subspace $\mathscr{X}_c$ is defined in the next theorem.

Theorem 2.5

The necessary and sufficient condition for the state $\mathbf{x}_0$ of the system (2.1) to be controllable is

$$\mathbf{x}_0 \in \mathscr{R}(\mathscr{C}) \tag{2.27}$$

The controllable subspace $\mathscr{X}_c$ is written as

$$\mathscr{X}_c = \mathscr{R}(\mathscr{C}) \tag{2.28}$$

where $\mathscr{R}(\mathscr{C})$ is the range space of the controllability matrix $\mathscr{C}$.

$$\mathscr{R}(\mathscr{C}) = \{\mathbf{x} \mid \mathbf{x} = \mathscr{C}\mathbf{u},\ ^\forall\mathbf{u} \in R^{mn}\}$$

Before the proof of this theorem an example is given.

Example 2.4 The controllable subspace of Example 2.1 is given by

$$\mathscr{R}(\mathscr{C}) = \mathscr{R}\begin{pmatrix} 1 & 0 & -1 \\ 1 & 1 & -3 \\ 0 & 1 & -2 \end{pmatrix} = \mathscr{R}\begin{pmatrix} 1 & 0 \\ 1 & 1 \\ 0 & 1 \end{pmatrix}$$

$$= \left\{ \alpha_1 \begin{pmatrix} 1 \\ 1 \\ 0 \end{pmatrix} + \alpha_2 \begin{pmatrix} 0 \\ 1 \\ 1 \end{pmatrix} \Bigg|\ \alpha_1, \alpha_2 \text{ are arbitrary real numbers} \right\}$$

†This is interpreted as the set $\mathscr{X}_c$ consists of $\mathbf{x}_0$ characterized by

$$\mathbf{x}_0 = -\int_0^t e^{-A\tau} B\mathbf{u}(\tau)\, d\tau$$

for any input.

The range space of a matrix is the subspace whose basis is formed by a linear combination of the independent column vectors of the matrix. Since here rank $\mathscr{C}$ is two we can take any two of the three column vectors as the basis vectors. Thus $\mathscr{R}(\mathscr{C})$ is a two-dimensional hyperplane, a subspace, of the three-dimensional space of A.

Proof of Theorem 2.5. By expanding $e^{-A\tau}$ (2.26) can be rewritten as

$$\int_0^t e^{-A\tau}B\mathbf{u}(\tau)\,d\tau = \int_0^t \left(I - A\tau + \frac{A^2}{2}\tau^2 - \frac{A^3}{3!}\tau^3 + \cdots\right)B\mathbf{u}(\tau)\,d\tau$$

$$= [B, AB, A^2B, \ldots]\begin{pmatrix} \int_0^t \mathbf{u}(\tau)\,d\tau \\ \int_0^t -\tau\mathbf{u}(\tau)\,d\tau \\ \int_0^t \frac{\tau^2}{2}\mathbf{u}(\tau)\,d\tau \\ \vdots \end{pmatrix}$$

where $\int_0^t \mathbf{u}(\tau)\,d\tau$, $\int_0^t \tau\mathbf{u}(\tau)\,d\tau$, $\int_0^t \tau^2\mathbf{u}(\tau)\,d\tau$, ... can be made arbitrary by an appropriate choice $\mathbf{u}(\tau)$. Therefore the controllability subspace $\mathscr{X}_c$ is represented by

$$\mathscr{X}_c = \mathscr{R}[B, AB, A^2B, \ldots]$$

From the Cayley-Hamilton theorem, $A^{n+i}B$ is given by a linear combination of $A^iB, A^{i+1}B, \ldots, A^{n+i-1}B$, and by repeating this procedure, $A^{n+k}B$ $(k > 0)$ can be represented by a linear combination of B, AB-, ..., $A^{n-1}B$. Therefore the controllable state in $\mathscr{X}_c$

$$\mathbf{x} = \sum_{i=0}^{\infty} A^iB\mathbf{u}_i$$

can be rewritten as

$$\mathbf{x} = \sum_{i=0}^{n-1} A^iB\mathbf{u}_i.$$

Thus

$$\mathscr{X}_c = \mathscr{R}[B, AB, \ldots]$$
$$= \mathscr{R}[B, AB, \ldots, A^{n-1}B]$$

and (2.28) holds.

For a system which is not completely controllable with rank $\mathscr{C} = n_c < n$ and with $B = [\mathbf{b}_1, \ldots, \mathbf{b}_m]$, the basis of $\mathscr{X}_c$, $\mathbf{v}_1, \ldots, \mathbf{v}_{n_c}$, can be obtained by choosing linearly independent vectors from $B, AB, \ldots, A^{n-1}B$. For this case, any element in the controllable subspace is given by

$$\mathbf{x}_c = \sum_{i=1}^{n_c} \mathbf{v}_i \bar{x}_i \qquad (\bar{x}_i \text{ is scalar})$$

In addition to $\mathbf{v}_1, \ldots$ to $\mathbf{v}_{n_c}$, linearly independent vectors $\mathbf{v}_{n_c+1}, \ldots$ to $\mathbf{v}_n$ can be introduced so that $\mathbf{v}_1, \ldots$ to $\mathbf{v}_n$ forms a basis for the n-dimensional vector space. Then any element $\mathbf{x}$ in the n-dimensional vector space can be represented by

$$\begin{aligned} \mathbf{x} &= \sum_{i=1}^{n} \mathbf{v}_i \bar{x}_i, \\ &= [\mathbf{v}_1, \mathbf{v}_2, \ldots, \mathbf{v}_n] \begin{pmatrix} \bar{x}_1 \\ \bar{x}_2 \\ \vdots \\ \bar{x}_n \end{pmatrix} \\ &= [\mathbf{v}_1, \mathbf{v}_2, \ldots, \mathbf{v}_{n_c}] \begin{pmatrix} \bar{x}_1 \\ \vdots \\ \bar{x}_{n_c} \end{pmatrix} + [\mathbf{v}_{n_c+1}, \ldots, \mathbf{v}_n] \begin{pmatrix} \bar{x}_{n_c+1} \\ \vdots \\ \bar{x}_n \end{pmatrix} \end{aligned} \tag{2.29}$$

The first term of the right side is the part in the controllable subspace. Since

$$T = [\mathbf{v}_1, \ldots, \mathbf{v}_n]$$

is non-singular, using this transformation the original system may be found to be equivalent to the special system given in the following theorem.

Theorem 2.6

When the system (2.1) is not completely controllable, it is equivalent to the system represented by

$$\dot{\bar{\mathbf{x}}} = \begin{pmatrix} \bar{A}_c & \bar{A}_{12} \\ O & \bar{A}_{22} \end{pmatrix} \bar{\mathbf{x}} + \begin{pmatrix} \bar{B}_c \\ O \end{pmatrix} \mathbf{u} \tag{2.30a}$$

$$\mathbf{y} = (\bar{C}_1, \bar{C}_2) \bar{\mathbf{x}} \tag{2.30b}$$

where $(\bar{A}_c, \bar{B}_c)$ is controllable.

Before the proof of this theorem the following example is considered.

Example 2.5 Let us find the equivalent system in the form (2.30) for the system in Example 2.1. Since

$$\mathscr{C} = [\mathbf{b}, A\mathbf{b}, A^2\mathbf{b}] = \begin{pmatrix} 1 & 0 & -1 \\ 1 & 1 & -3 \\ 0 & 1 & -2 \end{pmatrix}$$

$$\operatorname{rank} \mathscr{C} = 2$$

and from (2.29)

$$\mathbf{v}_1 = \begin{pmatrix} 1 \\ 1 \\ 0 \end{pmatrix}, \quad \mathbf{v}_2 = \begin{pmatrix} 0 \\ 1 \\ 1 \end{pmatrix}.$$

Choosing $\mathbf{v}_3$ as

$$\mathbf{v}_3 = \begin{pmatrix} 0 \\ 0 \\ 1 \end{pmatrix}$$

so that

$$T = [\mathbf{v}_1, \mathbf{v}_2, \mathbf{v}_3] = \begin{pmatrix} 1 & 0 & 0 \\ 1 & 1 & 0 \\ 0 & 1 & 1 \end{pmatrix}$$

is non-singular. The transformation T for

$$\mathbf{x} = T\bar{\mathbf{x}}$$

gives an equivalent system given by

$$\dot{\bar{\mathbf{x}}} = \begin{pmatrix} 1 & 0 & 0 \\ 1 & 1 & 0 \\ 0 & 1 & 1 \end{pmatrix}^{-1} \begin{pmatrix} 0 & 0 & -1 \\ 1 & 0 & -3 \\ 0 & 1 & -3 \end{pmatrix} \begin{pmatrix} 1 & 0 & 0 \\ 1 & 1 & 0 \\ 0 & 1 & 1 \end{pmatrix} \bar{\mathbf{x}} + \begin{pmatrix} 1 & 0 & 0 \\ 1 & 1 & 0 \\ 0 & 1 & 1 \end{pmatrix}^{-1} \begin{pmatrix} 1 \\ 1 \\ 0 \end{pmatrix} u$$

$$= \left(\begin{array}{cc|c} 0 & -1 & -1 \\ 1 & -2 & -2 \\ \hline 0 & 0 & -1 \end{array}\right) \bar{\mathbf{x}} + \left(\begin{array}{c} 1 \\ 0 \\ \hline 0 \end{array}\right) u$$

which is in the form of (2.30), where $\left(\begin{bmatrix} 0 & -1 \\ 1 & -2 \end{bmatrix}, \begin{bmatrix} 1 \\ 0 \end{bmatrix} \right)$ is obviously controllable.

Proof of Theorem 2.6 Let $\operatorname{rank} \mathscr{C} = n_c$, and pick n_c linearly independent vectors $\mathbf{v}_1, \mathbf{v}_2, \ldots, \mathbf{v}_{n_c}$ from $B, AB, \ldots, A^{n-1}B$, then

$$[\mathbf{v}_1, \ldots, \mathbf{v}_{n_c}] = [B, AB, \ldots, A^{n-1}B]M$$

Let the characteristic equation of A be as (1.24), then using the Cayley-Hamilton theorem the above equation yields

$$A[\mathbf{v}_1, \ldots, \mathbf{v}_{n_c}] = [AB, A^2B, \ldots, A^nB]M$$

$$= [B, AB, \ldots, A^{n-1}B] \begin{pmatrix} O & O & & O & -\alpha_0 I \\ I & O & & \vdots & \\ O & I & & & \vdots \\ \vdots & O & \ddots & & \\ & \vdots & & O & \\ O & O & & I & -\alpha_{n-1}I \end{pmatrix} M$$

This relation implies

$$A\mathbf{v}_i \in \mathscr{R}(\mathscr{C}) \qquad (i = 1, \ldots, n_c) \tag{2.31}$$

which means

$$A\mathbf{v}_i = \sum_{j=1}^{n_c} \bar{a}_{ji}\mathbf{v}_j \qquad (i = 1, \ldots, n_c) \tag{2.32}$$

For the linearly independent vectors $\{\mathbf{v}_{n_c+1}, \ldots, \mathbf{v}_n\}$ such that $\mathbf{v}_1, \mathbf{v}_2, \ldots, \mathbf{v}_n$ form the basis of an n-dimensional vector space, $A\mathbf{v}_i$ $(i = n_c + 1, \ldots, n)$, can be written as

$$A\mathbf{v}_i = \sum_{j=1}^{n_c} \bar{a}_{ji}\mathbf{v}_j \qquad (i = n_c + 1, \ldots, n) \tag{2.33}$$

From (2.32) and (2.33)

$$A[\mathbf{v}_1, \mathbf{v}_2, \ldots, \mathbf{v}_n] = [\mathbf{v}_1, \ldots, \mathbf{v}_n] \begin{pmatrix} \bar{a}_{11} & \ldots & \bar{a}_{1n_c} & \bar{a}_{1n_c+1} & \ldots & \bar{a}_{1n} \\ \vdots & & \vdots & \vdots & & \vdots \\ \bar{a}_{n_c1} & \ldots & \bar{a}_{n_cn_c} & & & \\ 0 & \ldots & 0 & & & \\ \vdots & & \vdots & & & \\ 0 & \ldots & 0 & \bar{a}_{nn_c+1} & \ldots & \bar{a}_{nn} \end{pmatrix}$$

On the other hand,

$$\mathbf{b}_i \in \mathscr{R}(\mathscr{C}) \qquad (i = 1, \ldots, m)$$

therefore

$$B = [\mathbf{v}_1, \ldots \mathbf{v}_{n_c}, \ldots, \mathbf{v}_n] \begin{pmatrix} \bar{b}_{11} & \ldots & \bar{b}_{1m} \\ \vdots & & \\ \bar{b}_{n_c1} & \ldots & \bar{b}_{n_cm} \\ 0 & \ldots & 0 \\ \vdots & & \vdots \\ 0 & \ldots & 0 \end{pmatrix}$$

From the above relations, by choosing

$$T = [\mathbf{v}_1, \ldots, \mathbf{v}_n]$$

the equivalent system is described by (2.30). The controllability matrix for (A, B) is written as

$$\mathscr{C} = (B, AB, \ldots, A^{n-1}B) = T\begin{pmatrix} \bar{B}_c, \bar{A}_c\bar{B}_c, \ldots, \bar{A}_c^{n-1}\bar{B}_c \\ O, \quad O, \quad \ldots, \quad O \end{pmatrix}$$

Using the Cayley-Hamilton theorem and rank $\mathscr{C} = n_c$, the following relation is derived.

$$\operatorname{rank}[\bar{B}_c, \bar{A}_c\bar{B}_c, \ldots, \bar{A}_c^{n_c-1}\bar{B}_c] = n_c$$

This equation indicates that $(\bar{A}_c, \bar{B}_c)$ is controllable.

In the proof of the theorem it was found that for any element $\mathbf{v}$ in the controllable subspace $\mathscr{R}(\mathscr{C})$, $A\mathbf{v}$ also belongs to the controllable subspace. This is written as

$$\mathbf{v} \in \mathscr{R}(\mathscr{C}) \Rightarrow A\mathbf{v} \in \mathscr{R}(\mathscr{C})$$

In general if for any element $\mathbf{v}$ of $\mathscr{X}_p$, $\mathbf{v} \in \mathscr{X}_p$, $A\mathbf{v}$ also belongs to $\mathscr{X}_p$, $A\mathbf{v} \in \mathscr{X}_p$, then $\mathscr{X}_p$ is said to be A-invariant, which is written as

$$A\mathscr{X}_p \subset \mathscr{X}_p \tag{2.34}$$

Thus the controllable subspace is A-invariant.

Theorem 2.6 shows that any system which is not completely controllable is equivalent to the decomposed system shown in Figure 2.2, where a free system $\dot{\mathbf{x}}_2 = A_{22}\mathbf{x}_2$ is not affected by the input and is uncontrollable. Therefore the system is decomposed into controllable and uncontrollable subsystems. Equivalently the whole state space of the system is composed of the controllable and uncontrollable subspaces.

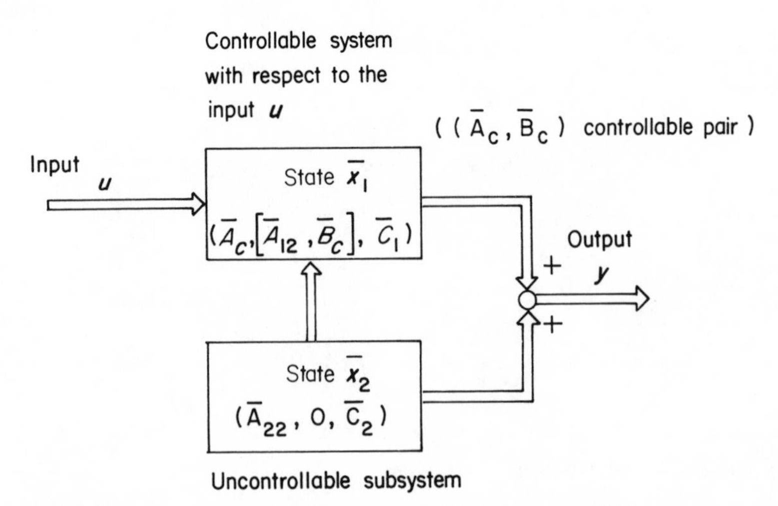

Figure 2.2 Decomposition into controllable and uncontrollable systems

In Example 1.6 it was shown that the transfer functions of equivalent systems are identical. Thus the transfer function matrix of the system (2.1) must be the same as that of (2.30). The transfer function of (2.30), $\bar{H}(s)$, is calculated from (1.127b) as

$$\bar{H}(s) = [\bar{C}_1, \bar{C}_2] \begin{pmatrix} sI - \bar{A}_c & -\bar{A}_{12} \\ O & sI - \bar{A}_{22} \end{pmatrix}^{-1} \begin{pmatrix} \bar{B}_c \\ O \end{pmatrix}$$

$$= [\bar{C}_1 \quad \bar{C}_2] \begin{pmatrix} (sI - \bar{A}_c)^{-1}, (sI - \bar{A}_c)^{-1}\bar{A}_{12}(sI - \bar{A}_{22})^{-1} \\ O \qquad (sI - \bar{A}_{22})^{-1} \end{pmatrix} \begin{pmatrix} \bar{B}_c \\ O \end{pmatrix}$$

$$= \bar{C}_1 (sI - \bar{A}_c)^{-1} \bar{B}_c \tag{2.35}$$

Therefore the input–output relationship for the system is only dependent on the controllable part of the system.

2.2.2 Unobservable Subspace and Kalman's Canonical Decomposition

Theorem 2.2 gives the condition for a system to be completely observable. However, as we have already seen in Example 2.3, there are systems which are not completely observable. In this section we consider the case where the time-invariant system (2.1) is not completely observable. Using the observability matrix

$$\mathcal{O} = [C^{\mathrm{T}}, A^{\mathrm{T}}C^{\mathrm{T}}, \ldots, (A^{\mathrm{T}})^{n-1}C^{\mathrm{T}}]$$

the n-dimensional state space R^n can be written as

$$R^n = \mathcal{R}(\mathcal{O}) \oplus \mathcal{N}(\mathcal{O}^{\mathrm{T}}). \tag{2.36}$$

Here $\mathcal{N}(\mathcal{O}^{\mathrm{T}})$ denotes the null space of $\mathcal{O}^{\mathrm{T}}$ defined as

$$\mathcal{N}(\mathcal{O}^{\mathrm{T}}) = \{\mathbf{x} \mid \mathcal{O}^{\mathrm{T}}\mathbf{x} = 0\}$$

which means that $\mathcal{N}(\mathcal{O}^{\mathrm{T}})$ is the set of elements $\mathbf{x}$ satisfying $\mathcal{O}^{\mathrm{T}}\mathbf{x} = 0$. The significance of (2.36) is that the basis for an n-dimensional vector space can be obtained from the range space of A and the null space of A^{T}. Since the rank of $\mathcal{O}$ is equal to the dimension of $\mathcal{R}(\mathcal{O})$ then for the completely observable system

$$\text{rank } \mathcal{O} = \dim \mathcal{R}(\mathcal{O}) = n$$

which indicates

$$\mathcal{N}(\mathcal{O}^{\mathrm{T}}) = \{0\}$$

However for a system which is not completely observable

$$\dim \mathcal{R}(\mathcal{O}) = n_0 < n$$

and from (2.36)

$$\dim \mathcal{N}(\mathcal{O}) = n_0 < n$$

and from (2.36)

$$\dim \mathcal{N}(\mathcal{O}^{\mathrm{T}}) = n - n_0$$

The dimension of $\mathcal{N}(\mathcal{O}^{\mathrm{T}})$ is called the nullity of $\mathcal{O}^{\mathrm{T}}$. The subspace $\mathcal{N}(\mathcal{O}^{\mathrm{T}})$ is called the unobservable subspace and any state in $\mathcal{N}(\mathcal{O}^{\mathrm{T}})$ is unobservable. When the initial state of the system (2.1) is unobservable, the signal $\boldsymbol{\eta}(t)$ given by (2.15) is always zero, since for any $\mathbf{x} \in \mathcal{N}(\mathcal{O}^{\mathrm{T}})$.

$$\begin{pmatrix} C \\ CA \\ \vdots \\ CA^{n-1} \end{pmatrix} \mathbf{x} = 0 \tag{2.37}$$

Using the relation $CA^i\mathbf{x} = 0 (i = 0, \ldots, n-1)$ and the Cayley-Hamilton theorem

$$\begin{aligned} \boldsymbol{\eta}(t) &= C\mathrm{e}^{At}\mathbf{x} \\ &= C\left(I + At + \frac{A^2}{2}t^2 + \frac{A^3}{3!}t^3 + \cdots\right)\mathbf{x} \\ &\equiv \mathbf{0} \end{aligned} \tag{2.38}$$

Conversely any initial state which gives $\boldsymbol{\eta}(t) \equiv \mathbf{0}$ is found from (2.38) to be unobservable.

Example 2.6 Let us check the observability of the system

$$\dot{\mathbf{x}} = \begin{pmatrix} 0 & 0 & -1 \\ 1 & 0 & -3 \\ 0 & 1 & -3 \end{pmatrix}\mathbf{x} + \begin{pmatrix} 1 \\ 1 \\ 0 \end{pmatrix}u$$

$$y = (0 \quad 1 \quad -2)\mathbf{x}$$

which has the same A and $\mathbf{b}$ as Example 2.1. The observability matrix of the system, $\mathcal{O}$, has rank given by

$$\operatorname{rank}\begin{pmatrix} 0 & 1 & -2 \\ 1 & -2 & 3 \\ -2 & 3 & -4 \end{pmatrix} = 2 < 3$$

and the system is not completely observable. The unobservable subspace $\mathcal{S}_{\bar{0}} = \mathcal{N}(\mathcal{O}^{\mathrm{T}})$ is

$$\mathcal{S}_{\bar{0}} = \mathcal{N}\begin{pmatrix} 0 & 1 & -2 \\ 1 & -2 & 3 \\ -2 & 3 & -4 \end{pmatrix} = \mathcal{N}\begin{pmatrix} 0 & 1 & -2 \\ 1 & -2 & 3 \end{pmatrix}$$

Noting that

$$\begin{pmatrix} 0 & 0 & 1 \\ 0 & 1 & -2 \\ 1 & -2 & 3 \end{pmatrix}^{-1} = \begin{pmatrix} 1 & 2 & 1 \\ 2 & 1 & 0 \\ 1 & 0 & 0 \end{pmatrix}$$

where the first row vector in the left side has been chosen to make the matrix non-singular, we can write this relationship

$$\begin{pmatrix} 0 & 0 & 1 \\ 0 & 1 & -2 \\ 1 & -2 & 3 \end{pmatrix} \begin{pmatrix} 1 & 2 & 1 \\ 2 & 1 & 0 \\ 1 & 0 & 0 \end{pmatrix} = \begin{pmatrix} 1 & 0 & 0 \\ 0 & 1 & 0 \\ 0 & 0 & 1 \end{pmatrix}$$

which yields

$$\begin{pmatrix} 1 \\ 2 \\ 1 \end{pmatrix} \in \mathcal{N} \begin{pmatrix} 0 & 1 & -2 \\ 1 & -2 & 3 \end{pmatrix}$$

An alternative way of interpreting this result is that the vector $(1,2,1)^{\mathrm{T}}$ is orthonormal to the vectors $(0,0,1)^{\mathrm{T}}$, $(0,1,-2)^{\mathrm{T}}$ and $(1,-2,3)^{\mathrm{T}}$. Thus dim $\mathcal{N}(\mathcal{C}^{\mathrm{T}}) = 1$ in this case and we can write

$$\mathcal{N} \begin{pmatrix} 0 & 1 & -2 \\ 1 & -2 & 3 \end{pmatrix} = \left\{ \alpha_1 \begin{pmatrix} 1 \\ 2 \\ 1 \end{pmatrix} \middle| \, \alpha_1 \text{ is an arbitrary real number} \right\}$$

which is written in general as

$$\left\{ \begin{pmatrix} 1 \\ 2 \\ 1 \end{pmatrix} \right\}.$$

The unobservable subspace $\mathcal{X}_{\bar{o}}$ of $\mathcal{N}(\mathcal{C}^{\mathrm{T}})$ can be shown to be A-invariant as follows.

For any $\mathbf{x} \in \mathcal{X}_{\bar{o}}$, there exists

$$\mathcal{C}^{\mathrm{T}} \mathbf{x} = 0$$

Let the characteristic equation of A be

$$\lambda^n + \alpha_{n-1}\lambda^{n-1} + \cdots + \alpha_0 = 0$$

then

$$\mathcal{C}^{\mathrm{T}} A \mathbf{x} = \begin{pmatrix} O, I, O & \ldots & O \\ O, O, I, O & \ldots & O \\ \vdots & & \vdots \\ O & \ldots & O, I \\ -\alpha_0 I & \ldots & -\alpha_{n-1} I \end{pmatrix} \begin{pmatrix} C \\ CA \\ \vdots \\ CA^{n-1} \end{pmatrix} \mathbf{x} = 0$$

so $A\mathbf{x} \in \mathcal{N}(\mathcal{C}^{\mathrm{T}})$ and the unobservable subspace $\mathcal{N}(\mathcal{C}^{\mathrm{T}})$ is A-invariant.

Theorem 2.7

When the time-invariant system (2.1) is not completely observable it is equivalent to a system of the form

$$\dot{\bar{\mathbf{x}}} = \begin{pmatrix} \bar{A}_{11} & O \\ \bar{A}_{21} & \bar{A}_{\bar{0}} \end{pmatrix} \bar{\mathbf{x}} + \begin{pmatrix} \bar{B}_1 \\ \bar{B}_2 \end{pmatrix} \mathbf{u} \tag{2.39a}$$

$$\mathbf{y} = (\bar{C}_0, O)\bar{\mathbf{x}} \tag{2.39b}$$

Proof. From the observability matrix

$$\mathcal{O}^{\mathsf{T}} = \begin{pmatrix} C \\ CA \\ \vdots \\ CA^{n-1} \end{pmatrix}$$

linearly independent row vectors $\mathbf{w}_1^{\mathsf{T}}$, $\mathbf{w}_2^{\mathsf{T}}$, $\mathbf{w}_{n_0}^{\mathsf{T}}$ may be chosen, where

$$n_0 = \text{rank } \mathcal{O}^{\mathsf{T}}.$$

Then $\mathbf{w}_{n_0+1}^{\mathsf{T}}, \dots, \mathbf{w}_n^{\mathsf{T}}$ are determined so that $\mathbf{w}_1^{\mathsf{T}}, \mathbf{w}_2^{\mathsf{T}}, \dots, \mathbf{w}_n^{\mathsf{T}}$ are mutually linearly independent from

$$\begin{pmatrix} \mathbf{w}_1^{\mathsf{T}} \\ \vdots \\ \mathbf{w}_n^{\mathsf{T}} \end{pmatrix}^{-1} = [\mathbf{v}_1, \dots, \mathbf{v}_{n_0}, \mathbf{v}_{n_0+1}, \dots, \mathbf{v}_n]$$

where $\mathbf{v}_{n_0+1}, \dots, \mathbf{v}_n$ are the basis of $\mathcal{N}(\mathcal{O}^{\mathsf{T}})$.

Since $\mathcal{N}(\mathcal{O}^{\mathsf{T}})$ is A-invariant,

$$A\mathbf{v}_i = \sum_{j=n_0+1}^{n} \bar{a}_{ji}\mathbf{v}_j \qquad (i = n_0 + 1, \dots, n)$$

and

$$A[\mathbf{v}_1, \mathbf{v}_2, \dots, \mathbf{v}_{n_0}, \mathbf{v}_{n_0+1}, \dots, \mathbf{v}_n] = [\mathbf{v}_1, \dots, \mathbf{v}_n] \begin{pmatrix} \bar{a}_{11} \dots \bar{a}_{1n_0} & 0 & \dots & 0 \\ \vdots \quad\quad \vdots & \vdots & & \vdots \\ & 0 & \dots & 0 \\ & \bar{a}_{n_0+1\,n_0+1} & \dots & \bar{a}_{n_0+1\,n} \\ & & & \vdots \\ \bar{a}_{n1} \dots \bar{a}_{nn_0} & & & \bar{a}_{nn} \end{pmatrix}$$

$$C = \left(\begin{array}{ccc|ccc} \bar{c}_{11} & \dots & \bar{c}_{1n_0} & 0 & \dots & 0 \\ \vdots & & \vdots & \vdots & & \vdots \\ \bar{c}_{p1} & \dots & \bar{c}_{pn_0} & 0 & \dots & 0 \end{array}\right) \begin{pmatrix} \mathbf{w}_1^{\mathsf{T}} \\ \mathbf{w}_2^{\mathsf{T}} \\ \vdots \\ \mathbf{w}_n^{\mathsf{T}} \end{pmatrix}$$

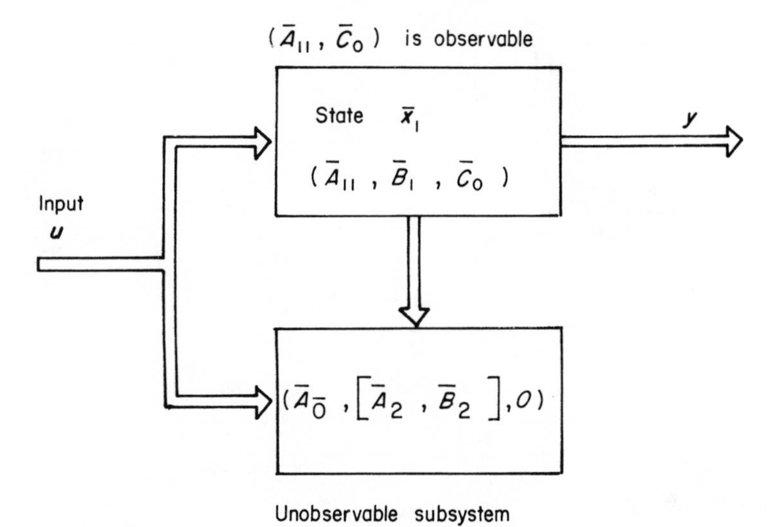

Figure 2.3 Decomposition into observable and unobservable systems

Therefore with the linear transformation

$$T = [\mathbf{v}_1, \ldots, \mathbf{v}_n]$$

the equivalent system in the form of (2.39) is obtained. The system structure of (2.39) is shown in Figure 2.3.

Example 2.7 Find the equivalent system of the form (2.39) for the system of Example 2.6. Since

$$\mathcal{N}(C^{\mathrm{T}}) = \begin{pmatrix} 1 \\ 2 \\ 1 \end{pmatrix}$$

T can be chosen as

$$T = \begin{pmatrix} 1 & 2 & 1 \\ 0 & 1 & 2 \\ 0 & 0 & 1 \end{pmatrix}.$$

Here we have changed the order of the columns in T from the matrix given in Example 2.6 to realize the system with the observable part in the upper block of the A matrix.

Using this transformation the equivalent system becomes

$$\bar{A} = T^{-1}AT = \begin{pmatrix} 1 & -2 & 3 \\ 0 & 1 & -2 \\ 0 & 0 & 1 \end{pmatrix} \begin{pmatrix} 0 & 0 & -1 \\ 1 & 0 & -3 \\ 0 & 1 & -3 \end{pmatrix} \begin{pmatrix} 1 & 2 & 1 \\ 0 & 1 & 2 \\ 0 & 0 & 1 \end{pmatrix}$$

$$= \left(\begin{array}{cc|c} -2 & -1 & 0 \\ 1 & 0 & 0 \\ \hline 0 & 1 & -1 \end{array}\right)$$

$$\bar{B}=T^{-1}B=\begin{pmatrix}-1\\ 1\\ \hdashline 0\end{pmatrix},\bar{C}=CT=(0 \quad 1 \ \vdots \ 0)$$

which has the form of (2.39).

The pair $(\bar{A}_{11},\bar{C}_0)$ given by (2.39) can easily be proved observable in a similar manner to Theorem 2.6.

The transfer function matrix of (2.39) $\bar{H}(s)$ is given by

$$\bar{H}(s)=[\bar{C}_0,O]\begin{pmatrix}sI-\bar{A}_{11} & O\\ -\bar{A}_{21} & sI-\bar{A}_{\bar{0}}\end{pmatrix}^{-1}\begin{pmatrix}\bar{B}_1\\ \bar{B}_2\end{pmatrix}$$

$$=[\bar{C}_0,O]\begin{pmatrix}(sI-\bar{A}_{11})^{-1} & O\\ (sI-\bar{A}_{\bar{0}})^{-1}\bar{A}_{21}(sI-\bar{A}_{11})^{-1} & (sI-\bar{A}_{\bar{0}})^{-1}\end{pmatrix}\begin{pmatrix}\bar{B}_1\\ \bar{B}_2\end{pmatrix}$$

$$=\bar{C}_0(sI-\bar{A}_{11})^{-1}\bar{B}_1 \tag{2.40}$$

which shows that the unobservable part of the system does not affect the input–output relationship.

Thus the state space can be decomposed into either controllable and uncontrollable or unobservable and observable subspaces. The generalization of Theorems 2.6 and 2.7 is given by Kalman's canonical decomposition theorem.

Theorem 2.8 (Kalman's canonical decomposition theorem)

The time invariant system of (2.1) is equivalent to a system of the following form.

$$\bar{A}=\begin{pmatrix}\bar{A}_{11} & \bar{A}_{12} & \bar{A}_{13} & \bar{A}_{14}\\ O & \bar{A}_{22} & O & \bar{A}_{24}\\ O & O & \bar{A}_{33} & \bar{A}_{34}\\ O & O & O & \bar{A}_{44}\end{pmatrix}\begin{matrix}\}n_1\\ \}n_2\\ \}n_3\\ \}n_4\end{matrix} \quad (\text{columns } n_1, n_2, n_3, n_4) \tag{2.41a}$$

$$\bar{B}=\begin{pmatrix}\bar{B}_1\\ \bar{B}_2\\ O\\ O\end{pmatrix}\begin{matrix}\}n_1\\ \}n_2\\ \}n_3\\ \}n_4\end{matrix} \tag{2.41b}$$

$$\bar{C}=(\underbrace{O}_{n_1},\underbrace{\bar{C}_2}_{n_2},\underbrace{O}_{n_3},\underbrace{\bar{C}_4}_{n_4}) \tag{2.41c}$$

where

$$\left.\begin{aligned} n_1&=\dim \mathscr{R}(\mathscr{C})\cap\mathscr{N}(\mathscr{O}^{\mathrm{T}})\\ n_2&=\dim \mathscr{R}(\mathscr{C})-n_1\\ n_3&=\dim \mathscr{N}(\mathscr{O}^{\mathrm{T}})-n_1\\ n_4&=n-n_1-n_2-n_3\end{aligned}\right\} \tag{2.42}$$

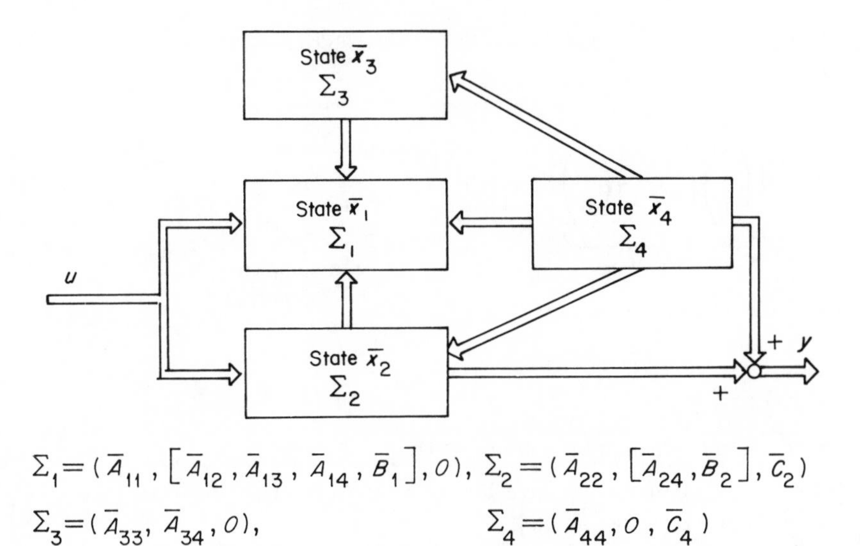

Figure 2.4 System decomposition

This theorem indicates that the state space can be decomposed into four subspaces, and the subsystems corresponding to the subspaces, shown in Figure 2.4, are denoted by Σ_1, Σ_2, Σ_3 and Σ_4, where Σ_1 is controllable but unobservable, Σ_2 is controllable and observable, Σ_3 is uncontrollable and unobservable, and Σ_4 is uncontrollable but observable.

Before the proof of the theorem an example is given.

Example 2.8 Find an equivalent system in the form of (2.41) for the system given in Example 2.6.

From Example 2.5,

$$\mathcal{R}(\mathcal{C}) = \left\{ \begin{pmatrix} 1 \\ 1 \\ 0 \end{pmatrix}, \begin{pmatrix} 0 \\ 1 \\ 1 \end{pmatrix} \right\}$$

and from Example 2.6,

$$\mathcal{N}(\mathcal{O}^{\mathrm{T}}) = \left\{ \begin{pmatrix} 1 \\ 2 \\ 1 \end{pmatrix} \right\}$$

Therefore $\mathcal{N}(\mathcal{O}^{\mathrm{T}}) \supset \mathcal{R}(\mathcal{C})$ and

$$\mathcal{R}(\mathcal{C}) \cap \mathcal{N}(\mathcal{O}^{\mathrm{T}}) = \left\{ \begin{pmatrix} 1 \\ 2 \\ 1 \end{pmatrix} \right\}; \text{ giving } n_1 = 1$$

Let

$$\mathcal{R}(\mathcal{C}) = \left\{ \begin{pmatrix} 1 \\ 2 \\ 1 \end{pmatrix} \right\} \oplus \left\{ \begin{pmatrix} 1 \\ 1 \\ 0 \end{pmatrix} \right\}$$

which indicates $n_2 = 1$, $n_3 = 0$, and $n_4 = 1$. The whole state space R^3 is represented by

$$R^3 = \overbrace{\left\{\begin{pmatrix}1\\2\\1\end{pmatrix}\right\}}^{\mathcal{N}(\mathcal{O}^{\mathrm{T}})} \underbrace{\oplus \left\{\begin{pmatrix}1\\1\\0\end{pmatrix}\right\}}_{\mathcal{R}(\mathcal{C})} \oplus \underbrace{\left\{\begin{pmatrix}0\\0\\1\end{pmatrix}\right\}}_{\text{arbitrary choice to complete } R^3}$$

Using the transformation

$$T = \begin{pmatrix} 1 & 1 & 0 \\ 2 & 1 & 0 \\ 1 & 0 & 1 \end{pmatrix}$$

the equivalent system in the form of (2.41) is

$$\bar{A} = T^{-1}AT = \begin{pmatrix} -1 & 1 & 0 \\ 2 & -1 & 0 \\ 1 & -1 & 1 \end{pmatrix}\begin{pmatrix} 0 & 0 & -1 \\ 1 & 0 & -3 \\ 0 & 1 & -3 \end{pmatrix}\begin{pmatrix} 1 & 1 & 0 \\ 2 & 1 & 0 \\ 1 & 0 & 1 \end{pmatrix} = \begin{pmatrix} -1 & 1 & -2 \\ 0 & -1 & 1 \\ 0 & 0 & -1 \end{pmatrix}$$

$$\bar{B} = T^{-1}B = \begin{pmatrix} -1 & 1 & 0 \\ 2 & -1 & 0 \\ 1 & -1 & 1 \end{pmatrix}\begin{pmatrix} 1 \\ 1 \\ 0 \end{pmatrix} = \begin{pmatrix} 0 \\ 1 \\ 0 \end{pmatrix}$$

$$\bar{C} = CT = (0 \quad 1 \quad -2)\begin{pmatrix} 1 & 1 & 0 \\ 2 & 1 & 0 \\ 1 & 0 & 1 \end{pmatrix} = (0 \quad 1 \quad -2)$$

where there is no $\bar{A}_{33}$ in $\bar{A}$ in this example. Except for $\mathcal{N}(\mathcal{O}^{\mathrm{T}}) \cap \mathcal{R}(\mathcal{C}) = \{[1, 2, 1]^{\mathrm{T}}\}$, the choice of the basis is not unique, so there exist many equivalent systems in the form of (2.41) besides the $(\bar{A}, \bar{B}, \bar{C})$ given here.

Proof of Theorem 2.8

Let

$$\mathcal{X}_{c\bar{o}} = \mathcal{R}(\mathcal{C}) \cap \mathcal{N}(\mathcal{O}^{\mathrm{T}})$$

then the controllable subspace $\mathcal{R}(\mathcal{C})$ can be expressed as the direct sum of the linear subspaces $\mathcal{X}_{c\bar{o}}$ and $\mathcal{X}_{co}$, so that

$$\mathcal{R}(\mathcal{C}) = \mathcal{X}_{c\bar{o}} \oplus \mathcal{X}_{co}$$

where the choice of $\mathcal{X}_{co}$ is not unique. Similarly a subspace $\mathcal{X}_{\bar{c}\bar{o}}$ is chosen so that the unobservable subspace $\mathcal{N}(\mathcal{O}^{\mathrm{T}})$ is expressed as the direct sum of $\mathcal{X}_{c\bar{o}}$ and $\mathcal{X}_{\bar{c}\bar{o}}$, that is

$$\mathcal{N}(\mathcal{O}^{\mathrm{T}}) = \mathcal{X}_{c\bar{o}} \oplus \mathcal{X}_{\bar{c}\bar{o}}$$

where the choice of $\mathcal{X}_{\bar{c}\bar{o}}$ is not unique. The whole state space is expressed as

$$R^n = \mathcal{X}_{c\bar{o}} \oplus \mathcal{X}_{co} \oplus \mathcal{X}_{\bar{c}\bar{o}} \oplus \mathcal{X}_{\bar{c}o} \tag{2.43}$$

where, as mentioned, the choices of $\mathscr{X}_{co}$, $\mathscr{X}_{\bar{c}\bar{o}}$ and $\mathscr{X}_{\bar{c}o}$ are not unique. Letting

$$\begin{aligned} &\dim \mathscr{X}_{c\bar{o}} = n_1 && \mathscr{X}_{c\bar{o}} = \{\mathbf{v}_1, \ldots, \mathbf{v}_{n_1}\} \\ &\dim \mathscr{X}_{co} = n_2 && \mathscr{X}_{co} = \{\mathbf{v}_{n_1+1}, \ldots, \mathbf{v}_{n_1+n_2}\} \\ &\dim \mathscr{X}_{\bar{c}\bar{o}} = n_3 && \mathscr{X}_{\bar{c}\bar{o}} = \{\mathbf{v}_{n_1+n_2+1}, \ldots, \mathbf{v}_{n_1+n_2+n_3}\} \\ &\dim \mathscr{X}_{\bar{c}o} = n_4 && \mathscr{X}_{\bar{c}o} = \{\mathbf{v}_{n_1+n_2+n_3+1}, \ldots, \mathbf{v}_{n}\} \end{aligned}$$

$\mathscr{X}_{c\bar{o}} = \mathscr{R}(\mathscr{C}) \cap \mathscr{N}(\mathscr{O}^{\mathrm{T}})$ is A-invariant, since $\mathscr{R}(\mathscr{C})$ and $\mathscr{N}(\mathscr{O}^{\mathrm{T}})$ are A-invariant, so for any $\mathbf{x} \in \mathscr{X}_{c\bar{o}}$, $A\mathbf{x} \in \mathscr{R}(\mathscr{C})$ and $A\mathbf{x} \in \mathscr{N}(\mathscr{O})^{\mathrm{T}}$ and therefore $A\mathbf{x} \in \mathscr{X}_{c\bar{o}}$. Thus

$$A\mathbf{v}_i = \sum_{j=1}^{n_1} \bar{a}_{ji}\mathbf{v}_j \qquad (i = 1, \ldots, n_1)$$

On the other hand $\mathscr{R}(\mathscr{C}) = \mathscr{X}_{c\bar{o}} \oplus \mathscr{X}_{co}$ is A-invariant, so

$$A\mathbf{v}_i = \sum_{j=1}^{n_1+n_2} \bar{a}_{ji}\mathbf{v}_i \qquad (i = n_1+1, \ldots, n_1+n_2)$$

Similarly $\mathscr{N}(\mathscr{O}^{\mathrm{T}}) = \mathscr{X}_{c\bar{o}} \oplus \mathscr{X}_{\bar{c}\bar{o}}$ is A-invariant

$$A\mathbf{v}_i = \sum_{j=1}^{n_1} \bar{a}_{ji}\mathbf{v}_j + \sum_{j=n_1+n_2+1}^{n_1+n_2+n_3} \bar{a}_{ji}\mathbf{v}_j \qquad (i = n_1+n_2+1, \ldots, n_1+n_2+n_3)$$

and $\mathscr{X}_{\bar{c}o}$ is not A-invariant

$$A\mathbf{v}_i = \sum_{j=1}^{n} \bar{a}_{ji}\mathbf{v}_j \qquad (i = n_1+n_2+n_3+1, \ldots, n)$$

From the above results (2.41a) is obtained. Equations (2.41b) and (2.41c) are derived from Theorems 2.6 and 2.7.

$\mathscr{R}(\mathscr{C}) \cap \mathscr{N}(\mathscr{O}^{\mathrm{T}})$ is calculated as follows: Letting

$$\begin{aligned} \mathscr{R}(\mathscr{C}) &= \{\mathbf{v}_1, \ldots, \mathbf{v}_{n_c}\} \\ \mathscr{N}(\mathscr{O}^{\mathrm{T}}) &= \{\mathbf{w}_1, \ldots, \mathbf{w}_{n_{\bar{o}}}\} \end{aligned}$$

the orthogonal space to $\mathscr{R}(\mathscr{C})$, $[\mathbf{v}_1, \ldots, \mathbf{v}_{n_c}]^{\perp}$, and that to $\mathscr{N}(\mathscr{O}^{\mathrm{T}})$, $\mathbf{w}_1, \ldots \mathbf{w}_{n_{\bar{o}}}]^{\perp}$ should be obtained. Then the intersection is given as follows: (Wonham)

$$\mathscr{R}(\mathscr{C}) \cap \mathscr{N}(\mathscr{O}^{\mathrm{T}}) = [[\mathbf{v}_1, \ldots, \mathbf{v}_{n_c}]^{\perp}, [\mathbf{w}_1, \ldots, \mathbf{w}_{n_{\bar{o}}}]^{\perp}]^{\perp}$$

The orthogonal space $[\mathbf{v}_1, \ldots, \mathbf{v}_{n_c}]^{\perp}$ is calculated from the inverse of $[\mathbf{v}_1, \ldots, \mathbf{v}_n]$ as follows, where $\mathbf{v}_{n_c+1}, \ldots, \mathbf{v}_n$ are determined so that the matrix is non-singular.

$$[\mathbf{v}_1, \ldots, \mathbf{v}_n]^{-1} = \begin{pmatrix} \mathbf{v}_1'^{\mathsf{T}} \\ \vdots \\ \mathbf{v}_{n_c}'^{\mathsf{T}} \\ \vdots \\ \mathbf{v}_n'^{\mathsf{T}} \end{pmatrix}$$

$$\mathbf{v}_i' \perp \{\mathbf{v}_1, \ldots, \mathbf{v}_{n_c}\}, \qquad i = n_c + 1, \ldots, n$$

therefore

$$[\mathbf{v}_1, \ldots, \mathbf{v}_{n_c}]^{\perp} = \{\mathbf{v}_{n_c+1}', \ldots, \mathbf{v}_n'\}$$
$$[\mathbf{w}_1, \ldots, \mathbf{w}_{n_{\bar{o}}}]^{\perp} = \{\mathbf{w}_{n_{\bar{o}}+1}', \ldots, \mathbf{w}_n'\}$$

and

$$[[\mathbf{v}_1, \ldots, \mathbf{v}_{n_c}]^{\perp}, [\mathbf{w}_1, \ldots, \mathbf{w}_{n_{\bar{o}}}]^{\perp}]^{\perp} = [\mathbf{v}_{n_c+1}', \ldots, \mathbf{v}_n', \mathbf{w}_{n_{\bar{o}}+1}', \ldots, \mathbf{w}_n']^{\perp}$$

Thus the intersection is calculated.

Example 2.9 Find

$$\mathscr{R}(A) \cap \mathscr{N}(A')$$

for the following matrices

$$A = \begin{pmatrix} 1 & 0 & -1 \\ 1 & 1 & -3 \\ 0 & 1 & -2 \end{pmatrix}, \quad A' = (0 \quad 1 \quad 0)$$

Now

$$\mathscr{R}(A) = \left\{\begin{pmatrix}1\\1\\0\end{pmatrix}, \begin{pmatrix}0\\1\\1\end{pmatrix}\right\}, \quad \mathscr{N}(A') = \left\{\begin{pmatrix}1\\0\\0\end{pmatrix}, \begin{pmatrix}1\\0\\1\end{pmatrix}\right\}$$

Choosing $\mathbf{v}_3$ as $(0 \quad 0 \quad 1)^{\mathsf{T}}$ gives

$$\begin{pmatrix} 1 & 0 & 0 \\ 1 & 1 & 0 \\ 0 & 1 & 1 \end{pmatrix}^{-1} = \begin{pmatrix} 1 & 0 & 0 \\ -1 & 1 & 0 \\ 1 & -1 & 1 \end{pmatrix}$$

(the third column being $\mathbf{v}_3$)

so that $(1, -1, 1)^{\mathsf{T}}$ is orthogonal to $(1, 1, 0)^{\mathsf{T}}$ and $(0, 1, 1)^{\mathsf{T}}$, that is

$$\begin{pmatrix} 1 & 0 \\ 1 & 1 \\ 0 & 1 \end{pmatrix}^{\perp} = \left\{\begin{pmatrix}1\\-1\\1\end{pmatrix}\right\}$$

Similarly we obtain

$$\begin{pmatrix} 1 & 0 \\ 0 & 0 \\ 0 & 1 \end{pmatrix}^{\perp} = \left\{\begin{pmatrix}0\\1\\0\end{pmatrix}\right\}$$

Using these two vectors we then find

$$\begin{pmatrix} 1 & 0 \\ -1 & 1 \\ 1 & 0 \end{pmatrix}^{\perp} = \left\{ \begin{pmatrix} -1 \\ 0 \\ 1 \end{pmatrix} \right\}$$

Therefore

$$\mathscr{R}(A) \cap \mathscr{N}(A') = \left\{ \begin{pmatrix} -1 \\ 0 \\ 1 \end{pmatrix} \right\}$$

Concerning the transfer function of (2.41) we have the following result.

Theorem 2.9

The transfer function of (2.41) is given by

$$\bar{H}(s) = \bar{C}_2(sI - \bar{A}_{22})^{-1}\bar{B}_2 \tag{2.44}$$

Proof. From (2.41)

$$\bar{H}(s) = (0 \quad \bar{C}_2 \quad 0 \quad \bar{C}_4) \begin{pmatrix} sI - \bar{A}_{11} & -\bar{A}_{12} & -\bar{A}_{13} & -\bar{A}_{14} \\ O & sI - \bar{A}_{22} & O & -\bar{A}_{24} \\ O & O & sI - \bar{A}_{33} & -\bar{A}_{34} \\ O & O & O & sI - \bar{A}_{44} \end{pmatrix}^{-1} \begin{pmatrix} \bar{B}_1 \\ \bar{B}_2 \\ O \\ O \end{pmatrix}$$

and from (1.127)

$$\begin{pmatrix} sI - \bar{A}_{11} & -\bar{A}_{12} & -\bar{A}_{13} & -\bar{A}_{14} \\ O & sI - \bar{A}_{22} & O & -\bar{A}_{24} \\ O & O & sI - \bar{A}_{33} & -\bar{A}_{34} \\ O & O & O & sI - \bar{A}_{44} \end{pmatrix}^{-1}$$

$$= \left(\begin{array}{cc|cc} (sI - \bar{A}_{11})^{-1} & (sI - \bar{A}_{11})^{-1}A_{12}(sI - \bar{A}_{22})^{-1} & \multicolumn{2}{c}{\multirow{2}{*}{X}} \\ O & (sI - \bar{A}_{22})^{-1} & & \\ \hline O & O & (sI - \bar{A}_{33})^{-1} & (sI - \bar{A}_{33})^{-1}\bar{A}_{34}(sI - \bar{A}_{44}) \\ O & O & O & (sI - \bar{A}_{44})^{-1} \end{array}\right)$$

where

$$X = \begin{pmatrix} (sI - \bar{A}_{11})^{-1}\bar{A}_{13}(sI - \bar{A}_{33})^{-1}, & (sI - A_{11})^{-1}\{\bar{A}_{12}(sI - \bar{A}_{22})^{-1}\bar{A}_{24} + \bar{A}_{13}(sI - \bar{A}_{33})^{-1} \\ & \bar{A}_{34} + \bar{A}_{14}\}(sI - \bar{A}_{44})^{-1} \\ O & (sI - \bar{A}_{22})^{-1}\bar{A}_{24}(sI - \bar{A}_{44})^{-1} \end{pmatrix}.$$

Substituting this expression for the inverse, (2.44) is derived.

$(\bar{A}_{22}, \bar{B}_2, \bar{C}_2)$ is completely controllable and completely observable, and only this part is found to affect the input–output relationship of the system.

2.2.3 Stability and Decomposition of the State Space

In the preceding sections, the state space has been decomposed from a controllability and observability viewpoint. In this section, the state space is decomposed from a stability viewpoint.

The initial state $\mathbf{x}_0$ of the time-invariant n-dimensional linear system

$$\dot{\mathbf{x}} = A\mathbf{x} \qquad \mathbf{x}(0) = \mathbf{x}_0 \tag{2.45}$$

is said to be unstable if its solution $\psi(t; \mathbf{x}_0, 0)$ is such that

$$\lim_{t\to\infty} \| \psi(t; \mathbf{x}_0, 0) \| = \infty \tag{2.46}$$

If there are no unstable states then the system is said to be stable. Further if for any initial state the solution satisfies

$$\lim_{t\to\infty} \| \psi(t; \mathbf{x}_0, 0) \| = 0, \ {}^\forall\mathbf{x}_0 \in R^n \tag{2.47}$$

the system of (2.45) is said to be asymptotically stable.

Letting the characteristic equation of A be

$$\det(sI - A) = s^n + \alpha_{n-1}s^{n-1} + \cdots + \alpha_0 = 0 \tag{2.48}$$

with roots $\lambda_1, \lambda_2, \ldots$ and λ_n, then there exists an initial state $\mathbf{v}_i$ to give the solution of (2.45) represented by

$$\psi(t; \mathbf{v}_i, 0) = e^{\lambda_i t}\mathbf{v}_i$$

If there exists such a solution then it satisfies (2.45). That is

$$\frac{d}{dt}(e^{\lambda_i t}\mathbf{v}_i) = \lambda_i e^{\lambda_i t}\mathbf{v}_i = Ae^{\lambda_i t}\mathbf{v}_i$$

Therefore the initial state satisfies

$$(\lambda_i I - A)\mathbf{v}_i = 0$$

which tells that $\mathbf{v}_i \in \mathcal{N}[(\lambda_i I - A)]$. Similarly for double roots λ_i, there exists an initial state $\mathbf{v}_i$ to give the solution

$$\psi(t; \mathbf{v}_i, 0) = \{e^{\lambda_i t}I - (\lambda_i I - A)te^{\lambda_i t}\}\mathbf{v}_i$$

If $\psi(t; \mathbf{v}_i, 0)$ is a solution,

$$\begin{aligned}\frac{\mathrm{d}}{\mathrm{d}t}&\{\mathrm{e}^{\lambda_i t}I-(\lambda_i I-A)t\mathrm{e}^{\lambda_i t}\}\mathbf{v}_i\\&=\{\lambda_i\mathrm{e}^{\lambda_i t}I-(\lambda_i I-A)\mathrm{e}^{\lambda_i t}-(\lambda_i I-A)\lambda_i t\mathrm{e}^{\lambda_i t}\}\mathbf{v}_i\\&=A\{\mathrm{e}^{\lambda_i t}I-(\lambda_i I-A)t\mathrm{e}^{\lambda_i t}\}\mathbf{v}_i\end{aligned}$$

which yields $(\lambda_i I - A)^2\mathbf{v}_i = 0$, therefore

$$\mathbf{v}_i \in \mathcal{N}\{(\lambda_i I - A)^2].$$

Generally for a p-tuple root λ_i, the initial state

$$\mathbf{v}_i \in \mathcal{N}[(\lambda_i I - A)^p]$$

gives the solution

$$\psi(t; \mathbf{v}_i, 0) = \sum_{k=0}^{p-1} \frac{(-1)^k}{k!}(\lambda_i I - A)^k t^k \mathrm{e}^{\lambda_i t}\mathbf{v}_i \tag{2.49}$$

Let C^+ denote the right half of the complex plane including the imaginary axis and let C^- be its complement, then the following polynomials can be defined for the characteristic roots $\lambda_1', \ldots, \lambda_k'$ of A with multiplicity $s_1, s_2, \ldots, s_k$ by

$$\varphi^+(s) = \prod_{i=1}^{l} (s-\lambda_i')^{s_i} \qquad \lambda_i \in C^+$$

$$\varphi^-(s) = \prod_{i=l+1}^{k} (s-\lambda_i')^{s_i} \qquad \lambda_i \in C^-$$

where λ_i' $(i = 1, \ldots, l)$ and λ_i' $(i = l+1, \cdots, k)$ are assumed in C^+ and C^- respectively. Since there is obviously no common factor for $\varphi^-(s)$ and $\varphi^+(s)$, there exist $h_1(s)$ and $h_2(s)$ such that

$$h_1(s)\varphi^+(s) + h_2(s)\varphi^-(s) = 1$$

Therefore any $\mathbf{x}$ in R^n satisfies

$$h_1(A)\varphi^+(A)\mathbf{x} + h_2(A)\varphi^-(A)\mathbf{x} = \mathbf{x} \tag{2.50}$$

From the Cayley-Hamilton theorem

$$\varphi^+(A)\varphi^-(A) = 0$$

which indicates

$$\mathbf{x} \in \mathcal{R}[h_1(A)\varphi^+(A)] \Rightarrow \varphi^-(A)\mathbf{x} = 0$$

therefore

$$\mathcal{R}[h_1(A)\varphi^+(A)] \subset \mathcal{N}[\varphi^-(A)]$$

On the other hand, for $\mathbf{x}_0 \in \mathcal{N}[\varphi^-(A)]$, $h_2(A)\varphi^-(A)\mathbf{x}_0 = 0$ and from (2.50)

$$\mathbf{x}_0 = h_1(A)\varphi^+(A)\mathbf{x}_0$$

therefore

$$\mathcal{N}[\varphi^-(A)] \subset \mathcal{R}[h_1(A)\varphi^+(A)]$$

Thus it is proved that

$$\mathcal{N}[\varphi^-(A)] = \mathcal{R}[h_1(A)\varphi^+(A)]$$

Similarly, it can be proved that

$$\mathcal{N}[\varphi^+(A)] = \mathcal{R}[h_2(A)\varphi^-(A)]$$

Therefore using (2.50) the whole state space can be written as

$$R^n = \mathcal{X}^+(A) \oplus \mathcal{X}^-(A) \tag{2.51}$$

where

$$\mathcal{X}^+(A) = \mathcal{N}[\varphi^+(A)]$$
$$\mathcal{X}^-(A) = \mathcal{N}[\varphi^-(A)]$$

$\mathcal{X}^+(A)$ includes $\mathcal{N}[(\lambda_i I - A)]$ with non-negative real part λ_i, so it is the set of all initial states which do not give asymptotically stable solutions, and $\mathcal{X}^-(A)$ is the set of all initial states which give asymptotically stable solutions. From the above discussion, the next theorem follows.

Theorem 2.10

The necessary and sufficient condition for the time-invariant system (2.45) to be asymptotically stable is that all its characteristic roots have negative real parts.

2.2.4 Lyapunov Function

In the previous section, the relation between stability and the real parts of the characteristic roots of A was discussed. In this section a Lyapunov function which can be used for the stability analysis of general systems, including nonlinear ones, is introduced for the stability analysis of a linear time-invariant system. Let the solution of a dynamical system be $\mathbf{x}(t)$, then a functional $V(\mathbf{x})$ which has first order partial derivatives with respect to $\mathbf{x}$, is real positive and has a derivative which is real and negative for all non-zero $\mathbf{x}$ is said to be a Lyapunov function, and a system for which a Lyapunov function can be constructed is asymptotically stable. Thus a Lyapunov function is useful for the stability analysis of general systems, but how to construct such a function is often a problem. For the linear system (2.45) a Lyapunov function can be constructed easily.

Theorem 2.11

A necessary and sufficient condition for the system (2.45) to be asymptotically stable is the existence of a positive symmetric matrix P satisfying

$$A^{\mathrm{T}}P + PA = -H^{\mathrm{T}}H \tag{2.52}$$

for H such that (A, H) is observable.

Proof For the positive functional defined by

$$V(\mathbf{x}) = \mathbf{x}^{\mathrm{T}}P\mathbf{x}$$

its derivative is

$$\begin{aligned}\dot{V}(\mathbf{x}) &= \dot{\mathbf{x}}^{\mathrm{T}}P\mathbf{x} + \mathbf{x}^{\mathrm{T}}P\dot{\mathbf{x}} \\ &= \mathbf{x}^{\mathrm{T}}(A^{\mathrm{T}}P + PA)\mathbf{x} \\ &= -\|H\mathbf{x}\|^2\end{aligned}$$

Since (A, H) is observable

$$= -\|He^{At}\mathbf{x}_0\|^2 < 0$$

so $V(\mathbf{x})$ is proved to be a Lyapunov function and the system is asymptotically stable. Necessity is proved as follows. Assume that the system is asymptotically stable, then the Grammian matrix in the form of

$$S \triangleq \int_0^{\infty} e^{A^{\mathrm{T}}\tau}H^{\mathrm{T}}He^{A\tau}\,d\tau \tag{2.53}$$

exists. Using the relation $e^{A^{\mathrm{T}}\tau}A^{\mathrm{T}} = A^{\mathrm{T}}e^{A^{\mathrm{T}}\tau}$,

$$\int_0^{\infty} \frac{d}{d\tau}(e^{A^{\mathrm{T}}\tau}H^{\mathrm{T}}He^{A\tau})\,d\tau = -H^{\mathrm{T}}H$$

$$\begin{aligned}&= \int_0^{\infty} (A^{\mathrm{T}}e^{A^{\mathrm{T}}\tau}H^{\mathrm{T}}He^{A\tau} + e^{A^{\mathrm{T}}\tau}H^{\mathrm{T}}He^{A\tau}A)\,d\tau \\ &= A^{\mathrm{T}}S + SA\end{aligned}$$

Letting $S = P$, (2.52) is obtained. Thus $\mathbf{x}^{\mathrm{T}}P\mathbf{x}$ is a Lyapunov function and (2.52) is said to be a Lyapunov equation.

PROBLEMS

P2.1 Find the controllable and unobservable subspaces of the following systems.

(1–1)
$$(A, \mathbf{b}, \mathbf{c}^{\mathrm{T}}) = \left(\begin{bmatrix} 0 & 0 & 0 \\ 1 & 0 & 1 \\ 1 & 0 & 0 \end{bmatrix}, \begin{bmatrix} 0 \\ 0 \\ 1 \end{bmatrix}, \begin{bmatrix} 0 & 1 & 0 \end{bmatrix} \right)$$

(1–2)
$$(A, \mathbf{b}, \mathbf{c}^{\mathrm{T}}) = \left(\begin{bmatrix} 0 & -2 \\ 1 & -3 \end{bmatrix}, \begin{bmatrix} 1 \\ 1 \end{bmatrix}, \begin{bmatrix} 0 & 1 \end{bmatrix} \right)$$

(1–3)
$$(A, \mathbf{b}, \mathbf{c}^{\mathrm{T}}) = \left(\begin{bmatrix} -2 & 0 & 1 \\ 0 & -2 & 1 \\ 0 & -3 & 2 \end{bmatrix}, \begin{bmatrix} 1 \\ 2 \\ 2 \end{bmatrix}, \begin{bmatrix} -1 & 1 & 0 \end{bmatrix} \right)$$

(1–4)
$$(A, B, \mathbf{c}^{\mathrm{T}}) = \left(\begin{bmatrix} -0.5 & -0.5 & 1.5 \\ 0.5 & -1.5 & 0.5 \\ 2 & -2 & 0 \end{bmatrix}, \begin{bmatrix} 1 & 1 \\ 1 & 2 \\ 0 & 1 \end{bmatrix}, \begin{bmatrix} 0 & 0 & 1 \end{bmatrix} \right)$$

(1–5)
$$(A, B, C) = \left(\begin{bmatrix} -1 & 3 & -1 \\ 3 & 1 & -4 \\ 2 & 3 & -4 \end{bmatrix}, \begin{bmatrix} 1 & 1 \\ 0 & 0 \\ 1 & 1 \end{bmatrix}, \begin{bmatrix} 1 & -1 & 0 \\ -1 & 0 & 1 \end{bmatrix} \right)$$

P2.2 For the systems in P2.1, find equivalent systems in the form of Kalman's canonical decomposition.

P2.3 Prove that the solution of the Lyapunov equation (2.52) for P is given as follows: Let A have distinct eigenvalues $\lambda_1, \ldots, \lambda_n$, then P is given by

$$P = [\mathbf{u}_1, \mathbf{u}_2, \ldots, \mathbf{u}_n]\,[\mathbf{v}_1, \mathbf{v}_2, \ldots, \mathbf{v}_n]^{-1} \tag{2.54}$$

where $[\mathbf{v}_i^{\mathrm{T}}, \mathbf{u}_i^{\mathrm{T}}]^{\mathrm{T}}$ is an eigenvector of

$$\begin{pmatrix} A & O \\ -H^{\mathrm{T}}H & -A^{\mathrm{T}} \end{pmatrix} \tag{2.55}$$

corresponding to the eigenvalue λ_i.

P2.4 The state space equation of a system is

$$\dot{\mathbf{x}} = A\mathbf{x} + B\mathbf{u}$$
$$\mathbf{y} = C\mathbf{x}$$

where

$$A = \begin{pmatrix} 2 & -6 & 0 \\ 6 & 2 & 0 \\ 0 & 0 & 1 \end{pmatrix}$$

check whether the system is observable if

(i) $C = (1 \quad 0 \quad 0)$ (ii) $C = (0 \quad 1 \quad 0)$

and (iii) $C = \begin{pmatrix} 1 & 0 & 0 \\ 0 & 0 & 1 \end{pmatrix}$

P2.5 Draw a block diagram for the system

$$(A, \mathbf{b}, \mathbf{c}^{\mathrm{T}}) = \left(\begin{bmatrix} -1 & 1 & a \\ 0 & -2 & 1 \\ 0 & 0 & -3 \end{bmatrix}, \begin{bmatrix} 0 \\ 0 \\ 1 \end{bmatrix}, [1 \quad 0 \quad 0] \right)$$

Discuss the controllability and observability of the system from the pole-zero pattern of the block diagram and check your results by manipulation on the matrices. In particular, find for what values of a the system may be uncontrollable or unobservable and determine the corresponding modes.

P2.6 Check whether the pair (A, B) is controllable for the following systems. If uncontrollable find the uncontrollable modes.

(i)
$$(A, \mathbf{b}) = \left(\begin{bmatrix} 0 & 1 \\ -2 & -3 \end{bmatrix}, \begin{bmatrix} 0 \\ 1 \end{bmatrix} \right)$$

(ii)
$$(A, B) = \left(\begin{bmatrix} 0 & 1 & 0 & 0 \\ 0 & 0 & 1 & 0 \\ 0 & 0 & 0 & 1 \\ -2 & -3 & -4 & -5 \end{bmatrix}, \begin{bmatrix} 1 & 0 \\ 0 & 1 \\ 1 & 0 \\ 0 & 1 \end{bmatrix} \right)$$

(iii)
$$(A, \mathbf{b}) = \left(\begin{bmatrix} 0 & 1 & 0 \\ 0 & 0 & 1 \\ 0 & -2 & -3 \end{bmatrix}, \begin{bmatrix} 1 \\ 0 \\ 0 \end{bmatrix} \right)$$

P2.7 Two systems S_1 and S_2 have state space representations (A_1, B_1, C_1, D_1) and (A_2, B_2, C_2, D_2), respectively, where A_i is $(n_i \times n_i)$, $B_i(n_i \times r_i)$, $C_i(m_i \times n_i)$ and $D_i(m_i \times r_i)$ for $i = 1, 2$. Show that

(i) If the two systems are connected in series (i.e. $m_1 = r_2$) then a necessary, but insufficient condition for the cascade combination to be controllable (observable) is that both S_1 and S_2 are controllable (observable).

(ii) If the two systems are connected in parallel (i.e. $r_1 = r_2$, $m_1 = m_2$) then a necessary and sufficient condition for the parallel combination to be controllable (observable) is that both S_1 and S_2 are controllable (observable).

3
CANONICAL FORMS AND MINIMAL REALIZATIONS

3.1 CANONICAL FORM

3.1.1 Canonical Form

This chapter is concerned with canonical forms. Canonical forms are important in studying the structure of classes of systems. A canonical form is a special form for representing a class of systems. There are many canonical forms, but in this chapter we first explain Luenberger's second controllable canonical form, and then the definition of the canonical form and its properties are discussed. We first consider a single input single output system.

Proposition

An n-dimensional single input system

$$\dot{\mathbf{x}} = A\mathbf{x} + \mathbf{b}u \tag{3.1}$$

has the equivalent system

$$\dot{\mathbf{x}}_c = \begin{pmatrix} 0 & & \\ \vdots & I_{n-1} & \\ 0 & & \\ -\alpha_0 & -\alpha_1 \ldots & -\alpha_{n-1} \end{pmatrix} \mathbf{x}_c + \begin{pmatrix} 0 \\ \vdots \\ 0 \\ 1 \end{pmatrix} u \tag{3.2}$$

if and only if $(A, \mathbf{b})$ is controllable, where

$$\det(Is - A) = s^n + \alpha_{n-1}s^{n-1} + \alpha_{n-2}s^{n-2} + \cdots + \alpha_0.$$

Proof If $(A, \mathbf{b})$ is controllable the controllability matrix

$$\mathscr{C} = [\mathbf{b}, A\mathbf{b}, \ldots, A^{n-1}\mathbf{b}]$$

is $n \times n$ and non-singular. Therefore $\mathscr{C}^{-1}$ exists. Let

$$\mathscr{C}^{-1} = L = \begin{pmatrix} \mathbf{l}_1^{\mathrm{T}} \\ \vdots \\ \mathbf{l}_n^{\mathrm{T}} \end{pmatrix}$$

and

$$T^{-1} = \begin{pmatrix} \mathbf{l}_n^{\mathrm{T}} \\ \mathbf{l}_n^{\mathrm{T}} A \\ \vdots \\ \mathbf{l}_n^{\mathrm{T}} A^{n-1} \end{pmatrix} \tag{3.3}$$

Now if T transforms A to the controllable form, A_c, then $T^{-1}A = A_c T^{-1}$ and this can be written

$$T^{-1}A = \begin{pmatrix} \mathbf{l}_n^{\mathrm{T}} A \\ \mathbf{l}_n^{\mathrm{T}} A^2 \\ \vdots \\ \mathbf{l}_n^{\mathrm{T}} A^n \end{pmatrix} = \begin{pmatrix} 0 & 1 & 0 & \dots & 0 \\ & & \cdot & & \\ & & & \cdot & \\ & & & & 1 \\ -\alpha_0 & & & \dots & -\alpha_{n-1} \end{pmatrix} \begin{pmatrix} \mathbf{l}_n^{\mathrm{T}} \\ \mathbf{l}_n^{\mathrm{T}} A \\ \vdots \\ \mathbf{l}_n^{\mathrm{T}} A^{n-1} \end{pmatrix}. \tag{3.4a}$$

This is true since the last row corresponds to the Cayley-Hamilton theorem result

$$A^n + \alpha_{n-1}A^{n-1} + \alpha_{n-2}A^{n-2} + \dots + \alpha_0 I = 0.$$

Also using $L\mathscr{C} = I$ we see that

$$T^{-1}\mathbf{b} = \begin{pmatrix} \mathbf{l}_n^{\mathrm{T}}\mathbf{b} \\ \mathbf{l}_n^{\mathrm{T}} A\mathbf{b} \\ \vdots \\ \mathbf{l}_n^{\mathrm{T}} A^{n-1}\mathbf{b} \end{pmatrix} = \begin{pmatrix} 0 \\ \vdots \\ 0 \\ 1 \end{pmatrix} \tag{3.4b}$$

and the sufficiency is proved.

Necessity. If (3.1) has the equivalent system of (3.2), then there exists a transformation matrix T satisfying

$$AT = T \begin{pmatrix} 0 & & \\ \vdots & I_{n-1} & \\ 0 & & \\ -\alpha_0 & \dots & -\alpha_{n-1} \end{pmatrix} \quad \text{and} \quad B = T \begin{pmatrix} 0 \\ \vdots \\ 0 \\ 1 \end{pmatrix}$$

Let the non-singular matrix

$$T = [\mathbf{t}_1, \dots, \mathbf{t}_n]$$

then from the above relations

$$\begin{aligned} \mathbf{t}_n &= \mathbf{b} \\ A\mathbf{t}_n &= \mathbf{t}_{n-1} - \alpha_{n-1}\mathbf{t}_n \quad \therefore \quad \mathbf{t}_{n-1} = A\mathbf{t}_n + \alpha_{n-1}\mathbf{b} \\ A\mathbf{t}_{n-1} &= \mathbf{t}_{n-2} - \alpha_{n-2}\mathbf{t}_n \quad \therefore \quad \mathbf{t}_{n-2} = A\mathbf{t}_{n-1} + \alpha_{n-2}\mathbf{b} \\ &\vdots \end{aligned}$$

which gives

$$\mathrm{T} = [A^{n-1}\mathbf{b} + \alpha_{n-1}A^{n-2}\mathbf{b} + \cdots + \alpha_1\mathbf{b}, \ldots, \mathbf{b}]$$

$$= [\mathbf{b}, A\mathbf{b}, \ldots, A^{n-1}\mathbf{b}] \begin{pmatrix} \alpha_1 & \alpha_2 & & 1 \\ \alpha_2 & \alpha_3 & & 0 \\ \vdots & & & \vdots \\ \alpha_{n-1} & 1 & & \\ 1 & 0 & \ldots & 0 \end{pmatrix} \tag{3.5}$$

since T is non-singular, $(A, \mathbf{b})$ is proved controllable. Q.E.D.

The system (3.2) is said to be in the controllable canonical form. Luenberger generalized this idea to multivariable systems.

Theorem 3.1

(Luenberger's second controllable canonical form). If an m input, p output, n-dimensional linear multivariable system (A, B, C) is controllable, then there exists a non-singular matrix T satisfying

$$A_c = T^{-1}AT, \quad B_c = T^{-1}B \tag{3.6}$$

where

$$A_c = \begin{pmatrix} \bar{A}_{11} & \ldots & \bar{A}_{1m} \\ \bar{A}_{21} & & \bar{A}_{2m} \\ \vdots & & \vdots \\ \bar{A}_{m1} & \ldots & \bar{A}_{mm} \end{pmatrix}, \; B_c = \begin{pmatrix} \bar{B}_1 \\ \bar{B}_2 \\ \vdots \\ \bar{B}_m \end{pmatrix} \tag{3.7}$$

and $\bar{A}_{ii} \in R^{\sigma_i \times \sigma_i}$, $\bar{A}_{ij} \in R^{\sigma_i \times \sigma_j}$, $\bar{B}_i \in R^{\sigma_i \times m}$ $(i = 1, \ldots, m, j = 1, \ldots, n)$ are given as follows:

$$\bar{A}_{ii} = \begin{pmatrix} 0 & & \\ \vdots & & I_{\sigma_i - 1} \\ 0 & & \\ -\alpha_{i\rho_{i-1}} & \ldots & -\alpha_{i(\rho_i - 1)} \end{pmatrix} \tag{3.8a}$$

$$\bar{A}_{ij} = \begin{pmatrix} 0 & \cdots & 0 \\ \vdots & & \vdots \\ 0 & \cdots & 0 \\ -\alpha_{i\rho_{j-1}} & \cdots & -\alpha_{i(\rho_j - 1)} \end{pmatrix}, i \neq j \tag{3.8b}$$

$$\bar{B}_i = \begin{pmatrix} 0 & \cdots & 0 & & \cdots & 0 \\ \vdots & & \vdots & & & 0 \\ & & 0 & & & 0 \\ 0 & \cdots & 1 & \beta_{(i+1)i} & \cdots & \beta_{mi} \end{pmatrix} \tag{3.8c}$$

with

$$\rho_i = \sum_{j=1}^{i} \sigma_j \qquad (i = 1, \ldots, m),\ \rho_0 = 0$$

and the σ_i are determined by the following procedure. Let $B = [\mathbf{b}_1, \mathbf{b}_2, \ldots, \mathbf{b}_m]$, then from $\{\mathbf{b}_1, \mathbf{b}_2, \ldots, \mathbf{b}_m, A\mathbf{b}_1, A\mathbf{b}_2, \ldots, A\mathbf{b}_m, A^2\mathbf{b}_1, \ldots\}$ n linearly independent column vectors are picked up sequentially. Let the set of these vectors be $\mathscr{S}$. If rank $B = m$, then

$$\mathscr{S} = \{\mathbf{b}_1, \mathbf{b}_2, \ldots, \mathbf{b}_m, A\mathbf{b}_1, \ldots\} \tag{3.9}$$

and σ_i is given by

$$\sigma_i = \max\{j \mid A^{j-1}\mathbf{b}_i \in \mathscr{S}\} \tag{3.10}$$

$\sigma_1, \sigma_2, \ldots, \sigma_m$ are called the controllability indices, and

$$\sigma = \max\{\sigma_1, \sigma_2, \ldots, \sigma_m\}$$

is known as the controllability index.

Proof Since (A, B) is controllable, $\mathscr{S}$ consists of n vectors. By rearranging the elements of $\mathscr{S}$, the following $n \times n$ matrix

$$S = [\mathbf{b}_1, A\mathbf{b}_1, A^2\mathbf{b}_1, \ldots, \mathbf{A}^{\sigma_1 - 1}\mathbf{b}_1, \mathbf{b}_2, \ldots, A^{\sigma_m - 1}\mathbf{b}_m] \tag{3.11}$$

is obtained where

$$\text{rank } B = m.$$

Since S is non-singular, the inverse matrix L exists, where

$$L = S^{-1} = \begin{pmatrix} \mathbf{l}_1^{\mathrm{T}} \\ \mathbf{l}_2^{\mathrm{T}} \\ \vdots \\ \mathbf{l}_{\rho_m}^{\mathrm{T}} \end{pmatrix} \tag{3.12}$$

Letting

$$T^{-1} = \begin{pmatrix} \mathbf{l}_{\rho_1}^{\mathsf{T}} \\ \mathbf{l}_{\rho_1}^{\mathsf{T}} A \\ \vdots \\ \mathbf{l}_{\rho_1}^{\mathsf{T}} A^{\sigma_1 - 1} \\ \mathbf{l}_{\rho_2}^{\mathsf{T}} \\ \vdots \\ \mathbf{l}_{\rho_m}^{\mathsf{T}} A^{\sigma_m} \end{pmatrix} \tag{3.13}$$

then

$$T^{-1}A = \begin{pmatrix} \mathbf{l}_{\rho_1}^{\mathsf{T}} A \\ \mathbf{l}_{\rho_1}^{\mathsf{T}} A^2 \\ \vdots \\ \mathbf{l}_{\rho_m}^{\mathsf{T}} A^{\sigma_m} \end{pmatrix} = \begin{pmatrix} \bar{A}_{11} \bar{A}_{12} & \dots & \bar{A}_{1m} \\ \bar{A}_{21} \bar{A}_{22} & & \bar{A}_{2m} \\ \vdots & & \vdots \\ \bar{A}_{m1} \bar{A}_{m2} & \dots & \bar{A}_{mm} \end{pmatrix} \begin{pmatrix} \mathbf{l}_{\rho_1}^{\mathsf{T}} \\ \mathbf{l}_{\rho_1}^{\mathsf{T}} A \\ \vdots \\ \mathbf{l}_{\rho_m}^{\mathsf{T}} A^{\sigma_m - 1} \end{pmatrix} \tag{3.14}$$

Since $\mathbf{l}_{\rho_i}^{\mathsf{T}} \cdot S = [0 \dots 0, 1, 0, \dots 0]$ as only the inner product of $\mathbf{l}_{\rho_i}$ and $A^{\sigma_i - 1}$ is equal to 1 and $\mathbf{l}_{\rho_i}^{\mathsf{T}}$ is orthogonal to $B = [\mathbf{b}_1, \dots, \mathbf{b}_m]$, $AB = [A\mathbf{b}_1, A\mathbf{b}_2, \dots, A\mathbf{b}_m]$, $A^{\sigma_i - 2}B = [A^{\sigma_i - 2}\mathbf{b}_1, \dots, A^{\sigma_i - 2}\mathbf{b}_m]$ and $A^{\sigma_i - 1}[\mathbf{b}_1, \mathbf{b}_2, \dots, \mathbf{b}_{i-1}]$. This gives

$$T^{-1}B = \begin{pmatrix} \mathbf{l}_{\rho_1}^{\mathsf{T}} B \\ \mathbf{l}_{\rho_1}^{\mathsf{T}} AB \\ \vdots \\ \vdots \\ \mathbf{l}_{\rho_1}^{\mathsf{T}} A^{\sigma_1 - 1} B \\ \vdots \end{pmatrix} = \begin{pmatrix} 0 & & 0 \\ \vdots & & \\ 0 & & \\ \vdots & & \\ 1 & \dots & X \\ 0 & \dots & 1 \end{pmatrix} \tag{3.15}$$

Q.E.D.

(A_c, B_c) are called the (second) controllable canonical form. When (A, C) is observable, $(A^{\mathsf{T}}, C^{\mathsf{T}})$ is controllable and the controllable canonical form of $(A^{\mathsf{T}}, C^{\mathsf{T}})$ can be obtained as $((A^{\mathsf{T}})_c, (C^{\mathsf{T}})_c)$. Letting

$$A_0 = (A^{\mathsf{T}})_c^{\mathsf{T}}, \; C_0 = (C^{\mathsf{T}})_c^{\mathsf{T}} \tag{3.16}$$

then (A_0, C_0) is said to be in Luenberger's second observable canonical form. This form is given by

$$A_0 = \begin{pmatrix} A_{11} & \dots & A_{2p} \\ A_{21} & \dots & A_{2p} \\ \vdots & & \vdots \\ A_{p1} & \dots & A_{pp} \end{pmatrix} \tag{3.17}$$

$$C_0 = [C_1, C_2, \dots, C_p] \tag{3.18}$$

where

$$\rho_j = \sum_{i=1}^{j} \sigma_i (j = 1, \ldots, p), \ \rho_0 = 0$$

Example 3.1 Find Luenberger's second controllable canonical form for the system

$$(A, B, C) = \left[\begin{pmatrix} 0 & 0 & 0 & 1 \\ 1 & 0 & 0 & -2 \\ -22 & -11 & -4 & 0 \\ -23 & -6 & 0 & -6 \end{pmatrix}, \begin{pmatrix} 0 & 0 \\ 0 & 0 \\ 0 & 1 \\ 1 & 3 \end{pmatrix}, \begin{pmatrix} 0 & 0 & 0 & 1 \\ 0 & 0 & 1 & 0 \end{pmatrix} \right]$$

The controllability matrix $\mathscr{C} = [B, AB, A^2B, A^3B]$ becomes

$$\mathscr{C} = \begin{pmatrix} 0 & 0 & 1 & 3 & -6 & -18 & 25 & 75 \\ 0 & 0 & -2 & -6 & 13 & 39 & -56 & -168 \\ 0 & 1 & 0 & -4 & 0 & 16 & -11 & -97 \\ 1 & 3 & -6 & -18 & 25 & 75 & -90 & -270 \end{pmatrix}$$

and the 1, 2, 3, and 5th column vectors are chosen sequentially as linearly independent vectors. So $\sigma_1 = 3$ and $\sigma_2 = 1$. By rearranging the vectors as $\mathbf{b}_1$, $A\mathbf{b}_1$, $A^2\mathbf{b}_1$, $\mathbf{b}_2$ we have

$$S = \begin{pmatrix} 0 & 1 & -6 & 0 \\ 0 & -2 & 13 & 0 \\ 0 & 0 & 0 & 1 \\ 1 & -6 & 25 & 3 \end{pmatrix}$$

where $B = [\mathbf{b}_1, \mathbf{b}_2]$.

The inverse of S is given by

$$L = S^{-1} = \begin{pmatrix} 28 & 11 & -3 & 1 \\ 13 & 6 & 0 & 0 \\ 2 & 1 & 0 & 0 \\ 0 & 0 & 1 & 0 \end{pmatrix} \begin{matrix} \\ \\ \leftarrow \sigma_1\text{th row} \\ \leftarrow \sigma_1 + \sigma_2\text{th row.} \end{matrix}$$

The inverse of the transformation matrix is constructed according to (3.13) as

$$T^{-1} = \begin{pmatrix} 2 & 1 & 0 & 0 \\ 1 & 0 & 0 & 0 \\ 0 & 0 & 0 & 1 \\ 0 & 0 & 1 & 0 \end{pmatrix} \begin{matrix} \leftarrow \ldots \mathbf{l}_{\rho_1}^{\mathrm{T}} \\ \leftarrow \ldots \mathbf{l}_{\rho_1}^{\mathrm{T}} A \\ \leftarrow \ldots \mathbf{l}_{\rho_1}^{\mathrm{T}} A^2 \\ \leftarrow \ldots \mathbf{l}_{\rho_2}^{\mathrm{T}} \end{matrix}$$

where $\rho_1 = \sigma_1$, $\rho_2 = \sigma_1 + \sigma_2$ and $\mathbf{l}_{\rho_i}^{\mathrm{T}}$ denotes the ρ_ith row vector of L. Therefore

$$T = \begin{pmatrix} 0 & 1 & 0 & 0 \\ 1 & -2 & 0 & 0 \\ 0 & 0 & 0 & 1 \\ 0 & 0 & 1 & 0 \end{pmatrix}$$

and the controllable canonical form (A_c, B_c, C_c) becomes

$$A_c = T^{-1}AT = \left(\begin{array}{cccc|c} 0 & 1 & 0 & 0 \\ 0 & 0 & 1 & 0 \\ -6 & -11 & -6 & 0 \\ \hline -11 & 0 & 0 & -4 \end{array}\right), \quad B_c = T^{-1}B = \left(\begin{array}{cc} 0 & 0 \\ 0 & 0 \\ 1 & 3 \\ \hline 0 & 1 \end{array}\right)$$

$$C_c = CT = \begin{pmatrix} 0 & 0 & 1 & 0 \\ 0 & 0 & 0 & 1 \end{pmatrix}$$

Example 3.2

Find the observable canonical form of the system

$$(A, B, C) = \left(\begin{bmatrix} 0 & 1 & 0 & 0 & 1 \\ 0 & 0 & 0 & 0 & -1 \\ 0 & 0 & 0 & 0 & 1 \\ 0 & 0 & 1 & 0 & -1 \\ 1 & 1 & 0 & 0 & 0 \end{bmatrix}, \begin{bmatrix} 1 & 0 \\ 0 & 0 \\ 0 & 1 \\ 0 & 1 \\ 0 & 0 \end{bmatrix}, \begin{bmatrix} 0 & 1 & 0 & 0 & 0 \\ 0 & 0 & -1 & 1 & 0 \end{bmatrix} \right)$$

The observability matrix $\mathcal{O} = [C^T, A^TC^T, \ldots, (A^T)^4C^T]$ is given by

$$\mathcal{O}^T = \begin{pmatrix} 0 & 0 & 0 & 0 & -1 & -2 & 0 & 1 & 0 & 0 \\ 1 & 0 & 0 & 0 & -1 & -2 & -1 & -1 & 0 & 1 \\ 0 & -1 & 0 & 1 & 0 & 0 & 0 & 0 & 0 & 0 \\ 0 & 1 & 0 & 0 & 0 & 0 & 0 & 0 & 0 & 0 \\ 0 & 0 & -1 & -2 & 0 & 1 & 0 & 0 & 1 & 2 \end{pmatrix}$$

and the 1, 2, 3, 4, 5 column vectors are sequentially chosen as linearly independent vectors. Letting

$$C = \begin{pmatrix} \mathbf{c}_1^T \\ \mathbf{c}_2^T \end{pmatrix}$$

S_0 is made up of $\mathbf{c}_1^T$, $\mathbf{c}_1^T A$, $\mathbf{c}_1^T A^2$, $\mathbf{c}_2^T$, $\mathbf{c}_2^T A$ and can be written

$$S_0 = \begin{pmatrix} 0 & 1 & 0 & 0 & 0 \\ 0 & 0 & 0 & 0 & -1 \\ -1 & -1 & 0 & 0 & 0 \\ 0 & 0 & -1 & 1 & 0 \\ 0 & 0 & 1 & 0 & -2 \end{pmatrix} \begin{matrix} \leftarrow \cdots \mathbf{c}_1^T \\ \leftarrow \cdots \mathbf{c}_1^T A \\ \leftarrow \cdots \mathbf{c}_1^T A^2 \\ \leftarrow \cdots \mathbf{c}_2^T \\ \leftarrow \cdots \mathbf{c}_2^T A \end{matrix}$$

where $\sigma_1 = 3$, $\sigma_2 = 2$. Let the inverse of S_0 be L_0, then

$$L_0 = \begin{pmatrix} -1 & 0 & -1 & 0 & 0 \\ 1 & 0 & 0 & 0 & 0 \\ 0 & -2 & 0 & 0 & 1 \\ 0 & -2 & 0 & 1 & 1 \\ 0 & -1 & 0 & 0 & 0 \end{pmatrix}$$
$$\qquad\qquad\quad \uparrow \qquad\quad \uparrow$$
$$\qquad\qquad\quad \rho_1 \qquad\quad \rho_2$$

Using the ρ_1 ($=\sigma_1$) and ρ_2 ($=\sigma_1+\sigma_2$)th column vectors $\mathbf{l}_1$, $\mathbf{l}_2$ then T is constructed as

$$T=\begin{pmatrix} -1 & 0 & -1 & 0 & 0\\ 0 & 0 & 1 & 0 & 0\\ 0 & 0 & -1 & 1 & 0\\ 0 & 0 & 1 & 1 & 1\\ 0 & -1 & 0 & 0 & 0 \end{pmatrix}$$
$$\qquad \mathbf{l}_{\rho_1} \quad A\mathbf{l}_{\rho_1} \quad A^2\mathbf{l}_{\rho_1} \quad \mathbf{l}_{\rho_2} \quad A\mathbf{l}_{\rho_2}$$

and T^{-1} is given by

$$T^{-1}=\begin{pmatrix} -1 & -1 & 0 & 0 & 0\\ 0 & 0 & 0 & 0 & -1\\ 0 & 1 & 0 & 0 & 0\\ 0 & 1 & 1 & 0 & 0\\ 0 & -2 & -1 & 1 & 0 \end{pmatrix}$$

Therefore the observable canonical form (A_0, B_0, C_0) is given by

$$A_0=T^{-1}AT=\left(\begin{array}{ccc|cc} 0 & 0 & -1 & 0 & 0\\ 1 & 0 & 0 & 0 & 0\\ 0 & 1 & 0 & 0 & 0\\ \hline 0 & 0 & 0 & 0 & 0\\ 0 & 0 & -1 & 1 & 0 \end{array}\right) \quad B_0=T^{-1}B=\left(\begin{array}{c|c} -1 & 0\\ 0 & 0\\ 0 & 0\\ \hline 0 & 1\\ 0 & 0 \end{array}\right)$$

$$C_0=CT=\left(\begin{array}{ccc|cc} 0 & 0 & 1 & 0 & 0\\ 0 & 0 & 2 & 0 & 1 \end{array}\right)$$

which has been constructed by the dual procedure to that for the controllable canonical form.

3.1.2 The Definition of Canonical Form and Ackermann's Procedure

In the previous section, the second controllable canonical form of Luenberger has been defined in (3.7) for a multi-input system. From this result we may consider that the following system

$$\dot{\mathbf{x}}=\left(\begin{array}{cc|c} 0 & 1 & 0\\ 1 & -1 & 1\\ \hline 1 & 2 & 2 \end{array}\right)\mathbf{x}+\left(\begin{array}{c|c} 0 & 0\\ 1 & 1\\ \hline 0 & 1 \end{array}\right)\mathbf{u} \tag{3.19}$$

is in controllable canonical form.

If, however, we again calculate the controllable canonical form using the procedure given in the previous section then the controllability matrix is given by

$$\mathscr{C}=[B, AB, A^2B]=\begin{pmatrix} 0 & 0 & 1 & 1 & -1 & 0\\ 1 & 1 & -1 & 0 & 4 & 5\\ 0 & 1 & 2 & 4 & 3 & 9 \end{pmatrix}.$$

The 1st, 2nd and 3rd column vectors are chosen sequentially as linearly independent. So S is defined as

$$S = [\mathbf{b}_1, A\mathbf{b}_1, \mathbf{b}_2] = \begin{pmatrix} 0 & 1 & 0 \\ 1 & -1 & 1 \\ 0 & 2 & 1 \end{pmatrix}$$

and yields

$$S^{-1} = L = \begin{pmatrix} \mathbf{l}_1^{\mathrm{T}} \\ \mathbf{l}_2^{\mathrm{T}} \\ \mathbf{l}_3^{\mathrm{T}} \end{pmatrix} = \begin{pmatrix} 3 & 1 & -1 \\ 1 & 0 & 0 \\ -2 & 0 & 1 \end{pmatrix} \begin{matrix} \\ \leftarrow \sigma_1 \\ \leftarrow \sigma_2 \end{matrix}$$

This gives the transformation matrix T as

$$T^{-1} = \begin{pmatrix} \mathbf{l}_2^{\mathrm{T}} \\ \mathbf{l}_2^{\mathrm{T}} A \\ \mathbf{l}_3^{\mathrm{T}} \end{pmatrix} = \begin{pmatrix} 1 & 0 & 0 \\ 0 & 1 & 0 \\ -2 & 0 & 1 \end{pmatrix}, \text{ and } T = \begin{pmatrix} 1 & 0 & 0 \\ 0 & 1 & 0 \\ 2 & 0 & 1 \end{pmatrix}$$

The controllable canonical form (A_c, B_c) is thus given by

$$A_c = T^{-1}AT = \begin{pmatrix} 0 & 1 & 0 \\ 3 & -1 & 1 \\ 5 & 0 & 2 \end{pmatrix} \tag{3.20a}$$

$$B_c = T^{-1}B = \begin{pmatrix} 0 & 0 \\ 1 & 1 \\ 0 & 1 \end{pmatrix} \tag{3.20b}$$

which is different from the original system (3.19). So we have to ask the question is (3.19) in the canonical form? What is the definition of the canonical form?

A definition of the canonical form has been given by many researchers such as Popov, Rissanen, Brunovsky, and Wang and Davison. The definition which Wang and Davison used is closely related to the equivalence relation, which is a relation satisfying reflexive, symmetric, and transitive properties. That is, let E be an equivalence relation on X, then xEx, $^\forall x \in X$; $xEy \Rightarrow yEx$, $^\forall x, y \in X$ and xEy, $yEz \Rightarrow xEz$, $^\forall x, y, z \in X$ define these properties.

Definition (invariant)

Let $\mathscr{X}$ be a set, and let E be an equivalence relation on X. If Γ is another set, a function $f: X \rightarrow \Gamma$ is said to be an invariant for E if $xEy \Rightarrow f(x) = f(y)$; and a complete invariant for E if $xEy \Leftrightarrow f(x) = f(y)$.

Definition (canonical form)

Let χ be a set, and let E be an equivalence relation on X. A map $\phi: X \to X$ is said to be a canonical map for E on X if

(1) $xE\phi(x), \; ^{\forall}x \in X$ (3.21a)

(2) $xEy \Leftrightarrow \phi(x) = \phi(y), \; ^{\forall}x, y \in X$ (3.21b)

The range of ϕ, denoted by $\mathscr{R}(\phi)$, is said to be a set of canonical forms for E on X.

Let X be a set of n-dimensional systems with m-inputs and p-outputs, then (A, B, C) is an element of X.

$$X = \{(A, B, C) \mid A \in R^{n \times n}, B \in R^{m \times n}, C \in R^{p \times n}\}$$

When (A, B, C) and $(\bar{A}, \bar{B}, \bar{C})$ are equivalent, that is, there exists a non-singular T satisfying $\bar{A} = T^{-1}AT$, $\bar{B} = T^{-1}B$ and $\bar{C} = CT$, we can denote $(A, B, C)E(\bar{A}, \bar{B}, \bar{C})$, since the relation is an equivalence relation. For the equivalence relation defined by

$$(A, B, C)E(\bar{A}, \bar{B}, \bar{C}) \Leftrightarrow (^{\exists}T\text{: non-singular})(\bar{A} = T^{-1}AT, \bar{B} = T^{-1}B, \bar{C} = CT) \quad (3.22)$$

there are many invariants. Controllability indices are invariants, since if $A^j\mathbf{b}_i$ is linearly independent of $\mathbf{b}_1, \mathbf{b}_2, \ldots, \mathbf{b}_m, A\mathbf{b}_1, \ldots, A^{j-1}\mathbf{b}_m, A^j\mathbf{b}_1, \ldots, A^j\mathbf{b}_{i-1}$, then $\bar{A}^j\bar{\mathbf{b}}_i = T^{-1}A^j\mathbf{b}_i$ is also linearly independent of $\bar{\mathbf{b}}_1, \bar{\mathbf{b}}_2, \ldots, \bar{\mathbf{b}}_m, \bar{A}\bar{\mathbf{b}}_1, \ldots, \bar{A}^{j-1}\bar{\mathbf{b}}_m, \bar{A}^j\bar{\mathbf{b}}_1, \ldots, \bar{A}^j\bar{\mathbf{b}}_{i-1}$. So if we use the following diagram, where a X (cross) is marked when A^j in the left of the row times $\mathbf{b}_i$ in the top of the column is linearly independent of the already

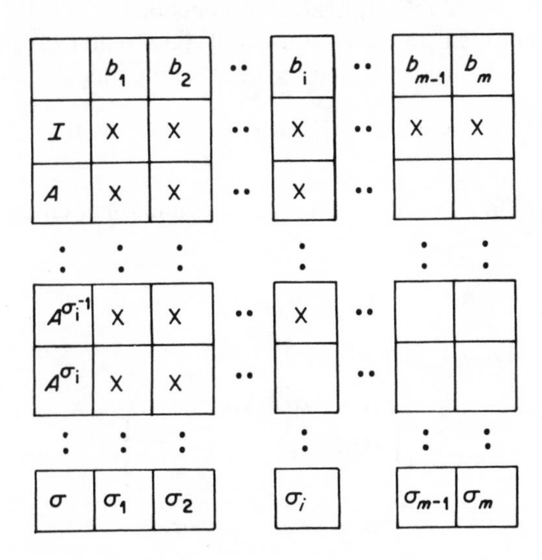

marked elements, then if the number of the cross marks in the ith column is σ_i,

$$A^{\sigma_i}\mathbf{b}_i = \sum_{j=1}^{i-1} \sum_{k=0}^{\sigma_i \wedge (\sigma_j - 1)} a_{ijk} A^k \mathbf{b}_j + \sum_{j=i}^{m} \sum_{k=0}^{(\sigma_i \wedge \sigma_j) - 1} a_{ijk} A^k \mathbf{b}_j \tag{3.23}$$

where $a \wedge b = \min(a, b)$. From the above relation we see that not only $\{\sigma_i\}$ but also $\{a_{ijk}\}$ are invariants on X. $\{\sigma_i\}$ is called a controllability index or Kronecker index.

Let the set of controllable systems be X_c defined as

$$X_c = \{(A, B, C) \mid (A, B, C) \in X, (A, B) \text{ is controllable}\}$$

The E defined in (3.22) is also an equivalence relation on X_c. For a system in X_c, the sum of Kronecker indices is n, and $\{\sigma_i\}$ plus $\{a_{ijk}\}$ are completely invariant for E on X_c, since

$$T = [\mathbf{b}_1, A\mathbf{b}_1, \ldots, A^{\sigma_1 - 1}\mathbf{b}_1, \mathbf{b}_2, \ldots, A^{\sigma_2 - 1}\mathbf{b}_2, \ldots, A^{\sigma_m - 1}\mathbf{b}_m]$$

gives a canonical form

$$\bar{A} = \left(\begin{array}{ccccc|c} 0 & 0 & \cdots & & a_{110} & 0 \\ 1 & 0 & & & a_{111} & \vdots \\ 0 & 1 & & & \vdots & \\ \vdots & \vdots & \ddots & & & \\ & & & 1 & a_{11\sigma_1 - 1} & 0 \\ & & & & a_{120} & 0 \\ & & & & \vdots & 1 \\ & & & & & \vdots \end{array}\right), \quad \bar{B} = \begin{pmatrix} 1 & 0 & \cdots & 0 \\ 0 & \vdots & & \vdots \\ \vdots & & & \\ & 1 & & \\ & 0 & & \\ & \vdots & & 1 \end{pmatrix}$$

for a class of controllable equivalent systems on X_c. The above canonical form is not convenient for general use and Ackermann gave a procedure to give the canonical form obtained by Luenberger's algorithm. He defined

$$\begin{aligned} \alpha_{ijk} &= a_{ijk} \text{ for } i = 1, \ldots, m, \ j = 1, \ldots, m, \ k = 0, \ldots, \min(\sigma_i, \sigma_j) - 1 \\ \beta_{ij} &= a_{ij\sigma_i} \end{aligned} \tag{3.24a}$$

where $\{\alpha_{ijk}\}$ and $\{\beta_{ij}\}$ are called α-parameters and β-parameters respectively. Let

$$\boldsymbol{\alpha}_{ij}^{\mathrm{T}} = [-\alpha_{ij0}, -\alpha_{ij1}, \ldots, -\alpha_{ij\bar{\sigma}_{ij}}, 0 \ldots 0] \tag{3.24b}$$

where $\bar{\sigma}_{ij} = \min(\sigma_i, \sigma_j) - 1$ and

$$K_{AB} = \begin{pmatrix} \boldsymbol{\alpha}_{11}^{\mathrm{T}} & \boldsymbol{\alpha}_{12}^{\mathrm{T}} & \cdots & \boldsymbol{\alpha}_{1m}^{\mathrm{T}} \\ \boldsymbol{\alpha}_{21}^{\mathrm{T}} & & & \\ \vdots & & & \\ \boldsymbol{\alpha}_{m1}^{\mathrm{T}} & \boldsymbol{\alpha}_{m2}^{\mathrm{T}} & \cdots & \boldsymbol{\alpha}_{mm}^{\mathrm{T}} \end{pmatrix} \tag{3.25a}$$

$$M_{AB} = \begin{pmatrix} 1 & -\beta_{21} & \cdots & -\beta_{m1} \\ 0 & 1 & & \vdots \\ \vdots & & & -\beta_{m(m-1)} \\ 0 & 0 & & 1 \end{pmatrix} \tag{3.25b}$$

then the canonical form is given by

$$A_c = A_e + B_e M_{AB}^{-1} K_{AB} \tag{3.26a}$$

$$B_c = B_e M_{AB}^{-1} \tag{3.26b}$$

where

$$A_e = \operatorname{diag}(A_{e1}, A_{e2}, \ldots, A_{em})$$

$$B_e = \operatorname{diag}(\mathbf{b}_{e1}, \ldots, \mathbf{b}_{em})$$

$$A_{ei} = \begin{pmatrix} 0 & 1 & & \\ & & \ddots & \\ & & & 1 \\ 0 & & \cdots & 0 \end{pmatrix} \quad \sigma_i \times \sigma_i \text{ matrix}$$

$$\mathbf{b}_{ei} = \begin{pmatrix} 0 \\ \vdots \\ 0 \\ 1 \end{pmatrix} \quad \sigma_i \times 1 \text{ matrix.}$$

Using Ackermann's procedure, the controllable canonical form of the system (3.19) is derived as follows:

$$\mathbf{b}_1 = \begin{pmatrix} 0 \\ 1 \\ 0 \end{pmatrix} \quad A\mathbf{b}_1 = \begin{pmatrix} 1 \\ -1 \\ 2 \end{pmatrix} \quad \mathbf{b}_2 = \begin{pmatrix} 0 \\ 1 \\ 1 \end{pmatrix}$$

$$A^2\mathbf{b}_1 = \begin{pmatrix} -1 \\ 4 \\ 3 \end{pmatrix} = -2\mathbf{b}_1 - A\mathbf{b}_1 + 5\mathbf{b}_2 = \alpha_{110}\mathbf{b}_1 + \alpha_{111}A\mathbf{b}_1 + \alpha_{120}\mathbf{b}_2$$

$$A\mathbf{b}_2 = \begin{pmatrix} 1 \\ 0 \\ 4 \end{pmatrix} = -\mathbf{b}_1 + A\mathbf{b}_1 + 2\mathbf{b}_2 = \alpha_{210}\mathbf{b}_1 + \beta_{21}A\mathbf{b}_1 + \alpha_{220}\mathbf{b}_2$$

Therefore

$$K_{AB} = \begin{pmatrix} \alpha_{110} & \alpha_{111} & \alpha_{210} \\ \alpha_{120} & 0 & \alpha_{220} \end{pmatrix} = \begin{pmatrix} -2 & -1 & -1 \\ 5 & 0 & 2 \end{pmatrix}$$

$$M_{AB} = \begin{pmatrix} 1 & -1 \\ 0 & 1 \end{pmatrix}$$

$$M_{AB}^{-1}K_{AB} = \begin{pmatrix} 1 & 1 \\ 0 & 1 \end{pmatrix}\begin{pmatrix} -2 & -1 & -1 \\ 5 & 0 & 2 \end{pmatrix} = \begin{pmatrix} 3 & -1 & 1 \\ 5 & 0 & 2 \end{pmatrix}$$

and the canonical form is

$$\dot{\mathbf{x}} = \begin{pmatrix} 0 & 1 & 0 \\ 3 & -1 & 1 \\ 5 & 0 & 2 \end{pmatrix} \mathbf{x} + \begin{pmatrix} 0 & 0 \\ 1 & 1 \\ 0 & 1 \end{pmatrix} \mathbf{u}$$

The relationship is proved by considering a simple example. For the controllable system,

$$\dot{\mathbf{x}} = \begin{pmatrix} 0 & 1 & 0 \\ a_{110} & a_{111} & a_{210} \\ a_{120} & 0 & a_{220} \end{pmatrix} \mathbf{x} + \begin{pmatrix} 0 & 0 \\ 1 & 0 \\ 0 & 1 \end{pmatrix} \mathbf{u}$$

The elements of the controllability matrix are

$$\mathbf{b}_1 = \begin{pmatrix} 0 \\ 1 \\ 0 \end{pmatrix} \qquad A\mathbf{b}_1 = \begin{pmatrix} 1 \\ a_{111} \\ 0 \end{pmatrix} \qquad A^2\mathbf{b}_1 = \begin{pmatrix} a_{111} \\ a_{110} + a_{111}^2 \\ a_{120} \end{pmatrix}$$

$$\mathbf{b}_2 = \begin{pmatrix} 0 \\ 0 \\ 1 \end{pmatrix} \qquad A\mathbf{b}_2 = \begin{pmatrix} 0 \\ a_{210} \\ a_{220} \end{pmatrix}$$

So that

$$A^2\mathbf{b}_1 = a_{110}\mathbf{b}_1 + a_{111}A\mathbf{b}_1 + a_{120}\mathbf{b}_2$$
$$A\mathbf{b}_2 = a_{210}\mathbf{b}_1 + a_{220}\mathbf{b}_2$$

Therefore in the case that all the β-parameters are zero, the coefficients of the linear combination giving $A^{\sigma_i}\mathbf{b}_i$ are the same as those in the A-matrix of the canonical form. However, if the β-parameters are not equal to zero such that

$$\dot{\mathbf{x}} = \begin{pmatrix} 0 & 1 & 0 \\ \tilde{\alpha}_{110} & \tilde{\alpha}_{111} & \tilde{\alpha}_{210} \\ \tilde{\alpha}_{120} & 0 & \tilde{\alpha}_{220} \end{pmatrix} \mathbf{x} + \begin{pmatrix} 0 & 0 \\ 1 & \beta_{21} \\ 0 & 1 \end{pmatrix} \mathbf{u}$$

$$\mathbf{b}_1 = \begin{pmatrix} 0 \\ 1 \\ 0 \end{pmatrix} \qquad A\mathbf{b}_1 = \begin{pmatrix} 1 \\ \tilde{\alpha}_{111} \\ 0 \end{pmatrix} \qquad A^2\mathbf{b}_1 = \begin{pmatrix} \tilde{\alpha}_{111} \\ \tilde{\alpha}_{110} + \tilde{\alpha}_{111}^2 \\ \tilde{\alpha}_{120} \end{pmatrix}$$

$$\mathbf{b}_2 = \begin{pmatrix} 0 \\ \beta_{21} \\ 1 \end{pmatrix} \qquad A\mathbf{b}_2 = \begin{pmatrix} \beta_{21} \\ \beta_{21}\tilde{\alpha}_{111} + \tilde{\alpha}_{210} \\ \tilde{\alpha}_{220} \end{pmatrix}$$

then

$$A^2\mathbf{b}_1 = \tilde{\alpha}_{110}\mathbf{b}_1 + \tilde{\alpha}_{111}A\mathbf{b}_1 + \tilde{\alpha}_{120}(\mathbf{b}_2 - \beta_{21}\mathbf{b}_1)$$

$$A(\mathbf{b}_2 - \beta_{21}\mathbf{b}_1) = \tilde{\alpha}_{210}\mathbf{b}_1 + \tilde{\alpha}_{220}(\mathbf{b}_2 - \beta_{21}\mathbf{b}_1).$$

This relationship shows that the coefficients of the following equations

$$A^2\mathbf{b}_1 = a_{110}\mathbf{b}_1 + a_{111}A\mathbf{b}_1 + a_{120}\mathbf{b}$$

$$A\mathbf{b}_2 = a_{210}\mathbf{b}_1 + a_{211}A\mathbf{b}_1 + a_{220}\mathbf{b}_2$$

satisfy $\beta_{21} = a_{211}$ and

$$\begin{pmatrix} 1 & -\beta_{21} \\ 0 & 1 \end{pmatrix}\begin{pmatrix} \tilde{\alpha}_{110} & \tilde{\alpha}_{111} & \tilde{\alpha}_{210} \\ \tilde{\alpha}_{120} & 0 & \tilde{\alpha}_{220} \end{pmatrix} = \begin{pmatrix} a_{110} & a_{111} & a_{210} \\ a_{120} & 0 & a_{220} \end{pmatrix},$$

and by generalizing the above example, Ackermann's procedure can be understood as follows:

From the above simple example, we will derive the controllable canonical form for a general multivariable system. $\tilde{B}$ is first defined as

$$\tilde{B} = BM_{AB}$$

where M_{AB} is

$$M_{AB} = \begin{pmatrix} 1 & -\beta_{21} & -\beta_{31} & \cdots & -\beta_{m1} \\ 0 & 1 & -\beta_{32} & \cdots & -\beta_{m2} \\ & & 1 & & \vdots \\ & 0 & & \ddots & -\beta_{mm-1} \\ & & & & 1 \end{pmatrix}$$

and the $\tilde{\alpha}$-parameters are defined as

$$\begin{pmatrix} \tilde{\alpha}_{i10} & \tilde{\alpha}_{i11} & \cdots & \tilde{\alpha}_{i1\sigma_i-1} \\ \tilde{\alpha}_{i20} & \tilde{\alpha}_{i21} & \cdots & \tilde{\alpha}_{i2\sigma_i-1} \\ \vdots & \vdots & & \vdots \\ \tilde{\alpha}_{im0} & \tilde{\alpha}_{im1} & \cdots & \tilde{\alpha}_{m\sigma_i-1} \end{pmatrix} = M_{AB}^{-1}\begin{pmatrix} a_{i10} & a_{i11} & \cdots & a_{i1\sigma_i-1} \\ a_{i20} & a_{i21} & \cdots & a_{i2\sigma_i-1} \\ \vdots & \vdots & & \vdots \\ a_{im0} & a_{im1} & \cdots & a_{im\sigma_i-1} \end{pmatrix}.$$

Then from (3.23)

$$A^{\sigma_i}\tilde{\mathbf{b}}_i = \tilde{B}\begin{pmatrix} \tilde{\alpha}_{i10} \\ \tilde{\alpha}_{i20} \\ \vdots \\ \tilde{\alpha}_{im0} \end{pmatrix} + A\tilde{B}\begin{pmatrix} \tilde{\alpha}_{i11} \\ \tilde{\alpha}_{i21} \\ \vdots \\ \tilde{\alpha}_{im1} \end{pmatrix} + \cdots + A^{\sigma_i-1}\tilde{B}\begin{pmatrix} \tilde{\alpha}_{i1\sigma_i} \\ \tilde{\alpha}_{i2\sigma_i} \\ \vdots \\ \tilde{\alpha}_{im\sigma_i} \end{pmatrix} \qquad (3.23)'$$

Let

$$T = [\mathbf{t}_{11}, \mathbf{t}_{12}, \ldots \mathbf{t}_{1\sigma_1}, \mathbf{t}_{21}, \mathbf{t}_{22}, \ldots \mathbf{t}_{2\sigma_2}, \ldots \mathbf{t}_{m\sigma_m}]$$

and

$$\begin{aligned}
\mathbf{t}_{i\sigma_i} &= \tilde{\mathbf{b}}_i \\
\mathbf{t}_{i\sigma_i-1} &= A\mathbf{t}_{i\sigma_i} - \tilde{B}\begin{pmatrix} \tilde{\alpha}_{i1\sigma_i-1} \\ \tilde{\alpha}_{i2\sigma_i-1} \\ \vdots \\ \tilde{\alpha}_{im\sigma_i-1} \end{pmatrix} \\
&\vdots \\
\mathbf{t}_{i1} &= A\mathbf{t}_{i2} - \tilde{B}\begin{pmatrix} \tilde{\alpha}_{i11} \\ \tilde{\alpha}_{i21} \\ \vdots \\ \tilde{\alpha}_{im1} \end{pmatrix},
\end{aligned}$$

then from (3.23)$'$

$$A\mathbf{t}_{i1} = \tilde{B}\begin{pmatrix} \tilde{\alpha}_{i10} \\ \tilde{\alpha}_{i20} \\ \vdots \\ \tilde{\alpha}_{im0} \end{pmatrix}$$

Using T, the following relations are derived

$$AT = T\begin{pmatrix}
0 & 1 & 0 & & & & & 0 \\
0 & 0 & 1 & & & & & \\
 & & & 1 & & & & \\
\tilde{\alpha}_{110} & \tilde{\alpha}_{111} & \cdots & \tilde{\alpha}_{11\sigma_1-1} & \tilde{\alpha}_{210} & & \cdots & \tilde{\alpha}_{m1\sigma_m-1} \\
0 & & & & 1 & & & \\
\vdots & & & & & & & \\
0 & & & & & & 1 & \\
\tilde{\alpha}_{120} & \tilde{\alpha}_{121} & \cdots & \tilde{\alpha}_{12\sigma_1-1} & \tilde{\alpha}_{220} & \cdots & \tilde{\alpha}_{22\sigma_2-1} & \cdots \\
0 & & & & & & & \\
\vdots & & & & & & & 1 \\
\tilde{\alpha}_{1m0} & \tilde{\alpha}_{1m1} & \cdots & \tilde{\alpha}_{1m\sigma_1-1} & \tilde{\alpha}_{2m0} & & \cdots & \tilde{\alpha}_{mm\sigma_m-1}
\end{pmatrix}$$

$$= TA_c$$

$$B = T\begin{pmatrix}
0 & & \cdots & 0 \\
\vdots & & & \\
0 & & & \\
1 & 0 & \cdots & 0 \\
0 & 0 & & \\
\vdots & \vdots & & \\
0 & 1 & \cdots & 0 \\
\vdots & & & \\
0 & & & 1
\end{pmatrix} M_{AB}^{-1} = TB_c$$

(A_c, B_c) are thus found in the canonical form given by Luenberger's algorithm.

3.1.3 Canonical Form and Input–Output Relation

The canonical form discussed has a close connection with the transfer function which represents the input–output relationship.

First the controllable canonical form $(A, \mathbf{b}, \mathbf{c}^\mathsf{T})$ with a single input and output is considered. We have

$$\dot{\mathbf{x}} = \begin{pmatrix} 0 & 1 & 0 & \dots & & 0 \\ \vdots & & & & & \vdots \\ & & & & & 0 \\ 0 & & & & 0 & 1 \\ -\alpha_0 & \dots & \dots & \dots & \dots & -\alpha_{n-1} \end{pmatrix} \mathbf{x} + \begin{pmatrix} 0 \\ \vdots \\ \\ 0 \\ 1 \end{pmatrix} u \tag{3.27a}$$

$$y = [\beta_0, \dots, \beta_{n-1}]\mathbf{x} \tag{3.27b}$$

and the transfer function of the system is

$$H(s) = \frac{\beta_{n-1}s^{n-1} + \cdots + \beta_0}{s^n + \alpha_{n-1}s^{n-1} + \cdots + \alpha_0} \tag{3.27c}$$

$$= \mathbf{c}^\mathsf{T} \begin{pmatrix} 1 \\ s \\ \vdots \\ s^{n-1} \end{pmatrix} \left[s^n - [-\alpha_0, \dots, -\alpha_{n-1}] \begin{pmatrix} 1 \\ s \\ \vdots \\ s^{n-1} \end{pmatrix} \right]^{-1} \tag{3.27c)'$$

which can be proved as follows: To show (3.27c)′ follows from (3.27c), we first prove

$$(sI - A)^{-1}\mathbf{b} = \begin{pmatrix} 1 \\ s \\ \vdots \\ s^{n-1} \end{pmatrix} \left[s^n - [-\alpha_0, \dots, -\alpha_{n-1}] \begin{pmatrix} 1 \\ s \\ \vdots \\ s^{n-1} \end{pmatrix} \right]^{-1} \tag{3.28}$$

Premultiplying by $(sI - A)$ and postmultiplying by

$$(s^n - [-\alpha_0, \dots, -\alpha_{n-1}])[1, s, \dots, s^{n-1}]^\mathsf{T}$$

(3.28) yields

$$\begin{pmatrix} 0 \\ \vdots \\ 0 \\ s^n + \alpha_{n-1}s^{n-1} + \cdots + \alpha_0 \end{pmatrix} = (sI - A) \begin{pmatrix} 1 \\ s \\ \dots \\ s^{n-1} \end{pmatrix} \tag{3.29}$$

which is obviously true. Therefore (3.28) holds and (3.27) follows from (1.23).

This relation between the transfer function and the canonical form of a single variable system can be generalized to a multivariable system by the structure theorem of Falb–Wolovich.

Theorem 3.2 (structure theorem)

Consider the second controllable canonical form of Luenberger (A_c, B_c, C_c) with full rank B represented by

$$A_c = \begin{pmatrix} A_{11}, A_{12}, A_{13} & \dots & A_{1m} \\ A_{21} & \dots & A_{2m} \\ \vdots & & \\ A_{m1} & \dots & A_{mm} \end{pmatrix}, \quad B_c = \begin{pmatrix} B_1 \\ \vdots \\ \\ B_m \end{pmatrix}, \quad C_c = [C_1, \dots, C_m] \tag{3.30}$$

$$A_{ii} = \begin{pmatrix} 0 & & I_{\sigma_i - 1} \\ -\alpha_{i\rho_{i-1}} & \dots & -\alpha_{i(\rho_i - 1)} \end{pmatrix}, \quad A_{ij} = \begin{pmatrix} & 0 & \\ -\alpha_{i\rho_{j-1}} & \dots & -\alpha_{i(\rho_j - 1)} \end{pmatrix} \tag{3.8a, b}$$

where $\rho_0 = 0$, $\rho_i = \Sigma_{j=1}^{i} \sigma_j$, and

$$B_j = \begin{pmatrix} & & 0 & & \\ 0 \dots \underset{\substack{\uparrow \\ j}}{1} & \beta_{jj} & & \dots & \beta_{j(m-1)} \end{pmatrix} \tag{3.8c}$$

The transfer matrix of the system is given by

$$C_c(sI - A_c)^{-1} B_c = C_c S(s)\, \delta_c^{-1}(s) \hat{B}_m \tag{3.31}$$

where

$$S(s) = \begin{pmatrix} S_1(s) \\ S_2(s) \\ \vdots \\ S_m(s) \end{pmatrix} \tag{3.32a}$$

$$S_i(s) = \begin{pmatrix} 0 & \dots & 0 & 1 & 0 & \dots & 0 \\ \vdots & & \vdots & s & & & \vdots \\ & & & \vdots & & & \\ 0 & \dots & 0 & \underset{\substack{\uparrow \\ i}}{s^{\sigma_i - 1}} & 0 & \dots & 0 \end{pmatrix} \tag{3.32b}$$

$$\delta_c(s) = \begin{pmatrix} s^{\sigma_1} & & 0 \\ & s^{\sigma_2} & \\ 0 & & \ddots \\ & & & s^{\sigma_m} \end{pmatrix} - \hat{A}_m S(s) \tag{3.33}$$

$$\hat{A}_m = \begin{pmatrix} -\alpha_{10} & \cdots & -\alpha_{1(n-1)} \\ \vdots & & \vdots \\ -\alpha_{m0} & \cdots & -\alpha_{m(n-1)} \end{pmatrix} \tag{3.34a}$$

$$\hat{B}_m = \begin{pmatrix} 1 & \beta_{11} & & \cdots \beta_{1(m-1)} \\ 0 & \ddots & & \\ \vdots & & & \\ & \ddots & 1 & \beta_{(m-1)(m-1)} \\ 0 & \cdots & 0 & 1 \end{pmatrix} \tag{3.34b}$$

The proof is given in the appendix of this chapter.

In the case where (A, B, C) is an observable canonical form a similar relation exists. Letting Luenberger's second observable canonical form of (A, B, C) be (A_0, B_0, C_0), then $(A_0^\mathsf{T}, C_0^\mathsf{T}, B_0^\mathsf{T})$ is Luenberger's second controllable canonical form; therefore the transfer matrix of the system from (3.31) is

$$B_0^\mathsf{T}(sI - A_0^\mathsf{T})^{-1}C_0^\mathsf{T}$$

and taking the transpose, the transfer matrix of (A_0, B_0, C_0) is given. Specifically

$$C_0(sI - A_0)^{-1}B_0 = \hat{C}_p\delta_0^{-1}(s)S^\mathsf{T}(s)B_0 \tag{3.35}$$

where $S(s)$ is defined similar to (3.32a,b), but by choosing $m = p$, and

$$\delta_0(s) = \begin{pmatrix} s^{\sigma_1} & & 0 \\ & s^{\sigma_2} & \\ & \ddots & \\ 0 & & s^{\sigma_p} \end{pmatrix} - S^\mathsf{T}(s)\begin{pmatrix} -\alpha_{01} & \cdots & -\alpha_{0p} \\ \vdots & & \\ -\alpha_{(n-1)1} & \cdots & -\alpha_{(n-1)p} \end{pmatrix} \tag{3.36}$$

$$\hat{C}_p = \begin{pmatrix} 1 & & \\ \gamma_{11} & 1 & 0 \\ \vdots & & \\ \gamma_{(p-1)1}, & \cdots & \gamma_{(p-1)(p-1)}, 1 \end{pmatrix} \tag{3.37}$$

Example 3.3 Find the transfer matrix of the system given in Example 3.1. Letting the controllable canonical form of (A, B, C) be (A_c, B_c, C_c), then by using the fact that equivalent systems have the same transfer matrix, there exists

$$H(s) = C(sI - A)^{-1}B = C_c(sI - A_c)^{-1}B_c$$

where

$$A_c = \begin{pmatrix} 0 & 1 & 0 & 0 \\ 0 & 0 & 1 & 0 \\ -6 & -11 & -6 & 0 \\ -11 & 0 & 0 & -4 \end{pmatrix}, \quad B_c = \begin{pmatrix} 0 & 0 \\ 0 & 0 \\ 1 & 3 \\ 0 & 1 \end{pmatrix}, \quad C_c = \begin{pmatrix} 0 & 0 & 1 & 0 \\ 0 & 0 & 0 & 1 \end{pmatrix}.$$

Using Theorem 3.2

$$H(s) = \begin{pmatrix} 0 & 0 & 1 & 0 \\ 0 & 0 & 0 & 1 \end{pmatrix} \begin{pmatrix} 1 & 0 \\ s & 0 \\ s^2 & 0 \\ 0 & 1 \end{pmatrix} \delta_c^{-1} \begin{pmatrix} 1 & 3 \\ 0 & 1 \end{pmatrix}$$

where

$$\delta_c(s) = \begin{pmatrix} s^3 & 0 \\ 0 & s \end{pmatrix} - \begin{pmatrix} -6 & -11 & -6 & 0 \\ -11 & 0 & 0 & -4 \end{pmatrix} \begin{pmatrix} 1 & 0 \\ s & 0 \\ s^2 & 0 \\ 0 & 1 \end{pmatrix}.$$

Therefore $H(s)$ is given by

$$H(s) = \begin{pmatrix} s^2 & 0 \\ 0 & 1 \end{pmatrix} \begin{pmatrix} s^3 + 6s^2 + 11s + 6 & 0 \\ 11 & s + 4 \end{pmatrix}^{-1} \begin{pmatrix} 1 & 3 \\ 0 & 1 \end{pmatrix}$$

$$= \frac{1}{s^4 + 10s^3 + 35s^2 + 50s + 24} \begin{pmatrix} s^3 + 4s^2 & 3s^3 + 12s^2 \\ -11 & s^3 + 6s^2 + 11s - 27 \end{pmatrix}$$

The result can be compared with that given by the algorithm of Faddeev (1.26) which yields

$$\Gamma_3 = I \qquad\qquad \alpha_3 = -\text{tr } A_c = 10$$

$$\Gamma_2 = A_c\Gamma_3 + \alpha_3 I = \begin{pmatrix} 10 & 1 & 0 & 0 \\ 0 & 10 & 1 & 0 \\ -6 & -11 & 4 & 0 \\ -11 & 0 & 0 & 6 \end{pmatrix}, \quad \alpha_2 = -\tfrac{1}{2}\text{tr} \begin{pmatrix} 0 & 10 & 1 & 0 \\ -6 & -11 & 4 & 0 \\ -24 & 60 & -35 & 0 \\ -66 & -11 & 0 & -24 \end{pmatrix} = 35$$

$$\Gamma_1 = A_c\Gamma_2 + \alpha_2 I = \begin{pmatrix} 35 & 10 & 1 & 0 \\ -6 & 24 & 4 & 0 \\ -24 & -50 & 0 & 0 \\ -66 & -11 & 0 & 11 \end{pmatrix}, \quad \alpha_1 = -\tfrac{1}{3}\text{tr} \begin{pmatrix} -6 & 24 & 4 & 0 \\ -24 & -50 & 0 & 0 \\ 0 & -24 & -50 & 0 \\ -121 & -66 & -11 & -44 \end{pmatrix} = 50$$

$$\Gamma_0 = A_c\Gamma_1 + \alpha_1 I = \begin{pmatrix} 44 & 24 & 4 & 0 \\ -24 & 0 & 0 & 0 \\ 0 & -24 & 0 & 0 \\ -121 & -66 & -11 & 6 \end{pmatrix}, \quad \alpha_0 = -\tfrac{1}{4}\text{tr} \begin{pmatrix} -24 & 0 & 0 & 0 \\ 0 & -24 & 0 & 0 \\ 0 & 0 & -24 & 0 \\ 0 & 0 & 0 & -24 \end{pmatrix} = 24$$

The transfer matrix is, therefore, given by

$$H(s) = \frac{1}{s^4 + 10s^3 + 35s^2 + 50s + 24} [C_cB_cs^3 + C_c\Gamma_2B_cs^2 + C_c\Gamma_1B_cs + C_c\Gamma_0B_c]$$

$$= \frac{1}{s^4 + 10s^3 + 35s^2 + 50s + 24}$$

$$\left\{\begin{pmatrix} 1 & 3 \\ 0 & 1 \end{pmatrix} s^3 + \begin{pmatrix} 4 & 12 \\ 0 & 6 \end{pmatrix} s^2 + \begin{pmatrix} 0 & 0 \\ 0 & 11 \end{pmatrix} s + \begin{pmatrix} 0 & 0 \\ -11 & -27 \end{pmatrix}\right\}$$

as before.

3.2 MINIMAL REALIZATION

The construction of a state space representation from an input–output relation such as the transfer function is called a realization. In Chapter 2, it was pointed out that the transfer function of a system is equal to that of its controllable and observable subsystem. From this fact it is therefore understandable that there can exist many state space representations with different orders which give the same transfer function. Among these possible state space descriptions the minimal dimension system is called a minimal realization. In this section, algorithms to obtain the minimal realization from a transfer function are given.

3.2.1 Dimension of a Minimal Realization

The transfer matrix $H(s)$ is a common representation for a system input–output relationship. For a multivariable system there are many forms for its representation. We give below several of these for strictly proper systems which by definition satisfy $\lim_{s\to\infty} H(s) = 0$.

(i)

$$H(s) = \begin{pmatrix} \dfrac{p_{11}(s)}{q_{11}(s)}, \dfrac{p_{12}(s)}{q_{12}(s)}, \ldots, \dfrac{p_{1m}(s)}{q_{1m}(s)} \\ \vdots \qquad\qquad \vdots \\ \dfrac{p_{p1}(s)}{q_{p1}(s)}, \dfrac{p_{p2}(s)}{q_{p2}(s)}, \ldots, \dfrac{p_{pm}(s)}{q_{pm}(s)} \end{pmatrix} \tag{3.38}$$

(ii)

$$H(s) = \frac{1}{q(s)} H_i(s) \tag{3.39}$$

where

$$q(s) = s^r + a_{r-1}s^{r-1} + \cdots + a_0, \qquad a_i \in R$$
$$H_i(s) = H_{r-1}s^{r-1} + \cdots + H_0, \qquad H_i \in R^{p\times m}$$

(iii) $$H(s) = N(s)D^{-1}(s) \tag{3.40a}$$
$$N(s) = N_{r-1}s^{r-1} + N_{r-2}s^{r-2} + \cdots + N_0, \qquad N_i \in R^{p\times m}$$
$$D(s) = D_r s^r + D_{r-1}s^{r-1} + \cdots + D_0, \qquad D_i \in R^{m\times m}$$

There is also a dual form to (3.40a) with

$$H(s) = N^{-1}(s)D(s) \tag{3.40b}$$
$$N(s) = N_r s^r + \cdots + N_0, \qquad N_i \in R^{p\times p}$$
$$D(s) = D_{r-1}s^{r-1} + \cdots + D_0, \qquad D_i \in R^{p\times m}$$

(iv) $$H(s) = H_0 s^{-1} + H_1 s^{-2} + H_2 s^{-3} + \cdots . \tag{3.41}$$

In this representation the coefficient matrices H_i of (3.41) are called Markov parameters.

If the transfer matrix of the system (A, B, C) is $H(s)$, that is,

$$\begin{aligned} H(s) &= C(sI - A)^{-1}B \\ &= CBs^{-1} + CABs^{-2} + \cdots \end{aligned}$$

than the Markov parameter H_i is related to (A, B, C) by

$$H_i = CA^iB, \qquad i = 0, 1, 2, \ldots \tag{3.42}$$

Therefore if there exists an A, B, C satisfying (3.42) then (A, B, C) is a realization of $H(s)$. Theorem 2.8 in Chapter 2, shows that the completely controllable and observable subsystem of (A, B, C) gives the same transfer matrix. There is, however, a question of whether there is a system with a lower order than that of the completely controllable and observable system. The next theorem answers this question.

Theorem 3.3

The realization of the system (A, B, C) from the given transfer function matrix $H(s)$, is a minimal realization if and only if it is completely controllable and observable.

Proof From Theorem 2.8, it is known that the minimal realization is controllable and observable. Therefore it will be proved that the controllable and observable system is a minimal realization. Let a controllable and observable nth order system (A, B, C) be a realization of $H(s)$ and assume that there exists a lower order system (A', B', C') with the dimension $n'(<n)$, then a Hankel matrix composed of the Markov parameters

$$\mathscr{H}_n = \begin{pmatrix} H_0, H_1, H_2 & \cdots & H_{n-1} \\ H_1, H_2, H_3 & \cdots & H_n \\ \vdots & & \\ H_{n-1}, H_n, H_{n+1} & \cdots & H_{2n-2} \end{pmatrix} \tag{3.43}$$

can be rewritten using the relation $H_i = CA^iB$ as

$$\mathscr{H}_n = \begin{pmatrix} C \\ CA \\ CA^2 \\ \vdots \\ CA^{n-1} \end{pmatrix} [B, AB, \ldots, A^{n-1}B]$$

By the assumption that (A, B) is controllable and (C, A) is observable,

$$\text{rank } \mathscr{H}_n = n \tag{3.44}$$

However $H_i = C'A'^iB'$ gives

$$\mathscr{H}_n = \begin{pmatrix} C' \\ C'A' \\ \vdots \\ C'A'^{n-1} \end{pmatrix} [B', A'B', \ldots, A'^{n-1}B']$$

Therefore

$$\text{rank}[(C')^{\mathsf{T}}, (C'A')^{\mathsf{T}}, \ldots, (C'A'^{n-1})^{\mathsf{T}}] \leqslant n'(<n)$$

$$\text{rank}[B', A'B', \ldots, A'^{n-1}B'] \leqslant n'(<n)$$

which yields

$$\text{rank } \mathscr{H}_n \leqslant n'(<n)$$

This result contradicts (3.44) and thus it is proved that the controllable and observable realization is minimal.

This theorem shows the uniqueness of the order of the minimal realization, and all minimal realizations for a given input–output relation are shown equivalent by the following theorem.

Theorem 3.4

If (A_1, B_1, C_1) and (A_2, B_2, C_2) are minimal realizations for the transfer matrix $H(s)$, then they are equivalent.

Proof It is given in the appendix.

3.2.2 Kalman's Algorithm for a Minimal Realization from a Transfer Matrix

There are many algorithms for obtaining a minimal realization from an input–output relationship. First, the algorithm of R. E. Kalman which is historically important is described.

When the transfer matrix $H(s)$ is considered, the least common multiple of all denominators of components in $H(s)$ is denoted by $\psi(s)$. Then $\psi(s)H(s)$ becomes a polynomial matrix and can be written in the Smith form as

$$\psi(s)H(s) = P(s)\begin{pmatrix} \nu_1(s) & & & & & \\ & \nu_2(s) & & & & \\ & & \ddots & & & \\ & & & 0 & & \\ & & & & \ddots & \\ & & & & & 0 \end{pmatrix} Q(s) \tag{3.45}$$

where $P(s)$ and $Q(s)$ are unimodular matrices and

$$\nu_1(s) \mid \nu_2(s) \mid \ldots$$

(i.e. $\nu_1(s)$ divides $\nu_2(s)$).

Dividing both sides of (3.45) by $P(s)$, the following MacMillan form is obtained for $H(s)$.

$$H(s) = P(s)\begin{pmatrix} \dfrac{\nu_1'(s)}{\mu_1(s)} & & & & 0 \\ & \dfrac{\nu_2'(s)}{\mu_2(s)} & & & \\ & & \ddots & & \\ 0 & & & \dfrac{\nu_r'(s)}{\mu_r(s)} & \\ & & & & \dfrac{\nu_{r+1}'(s)}{\mu_{r+1}(s)} \end{pmatrix} \tag{3.46}$$

where $\nu_i'(s)$ and $\mu_i(s)$ are obtained by cancelling the common factors of $\nu_i(s)$ and $\psi(s)$ so that they are coprime and satisfy

$$\nu_1'(s) \mid \nu_2'(s) \mid \ldots$$

and

$$\mu_r(s) \mid \mu_{r-1}(s) \mid \mu_{r-2}(s) \mid \ldots .$$

If $H(s)$ is a minimal realization of a strongly proper linear system, its dimension is $\Sigma_{i=1}^{r} \deg(\mu_i(s))$ and it is called the MacMillan degree.

Equation (3.46) can be rewritten as

$$\psi(s)H(s) = \sum_{i=1}^{r} \left\{ \mathbf{p}_i(s) \frac{\nu_i'(s)}{\mu_i(s)} \mathbf{q}_i^{\mathrm{T}}(s) \right\} \psi(s) \pmod{\psi(s)} \tag{3.47}$$

where $\mathbf{p}_i(s)$ is the ith column of $P(s)$ and $\mathbf{q}_i^{\mathrm{T}}(s)$ is the ith row of $Q(s)$. Let

$$A_i = \begin{pmatrix} 0 & 1 & & \\ & & \ddots & \\ & & & 1 \\ -a_{(n_i-1)i} & & \cdots & -a_{1i} \end{pmatrix} \qquad i = 1, \ldots, r \tag{3.48}$$

then the following relation exists

$$(sI - A_i)^{-1}\psi_i(s) \equiv$$

$$\begin{pmatrix} 1 \\ s \\ s^2 \\ \vdots \\ s^{n_i-1} \end{pmatrix} [s^{n_i-1} + a_{1i}s^{n_i-2} + \cdots + a_{(n_i-1)i}, \ldots, s + a_{1i,1}] \pmod{\psi_i(s)} \tag{3.49}$$

where

$$\psi_i(s) = s^{n_i} + a_{1i}s^{n_i-1} + \cdots + a_{n_i1}$$

Using the relation (3.49), $H(s)$ of (3.47) is realized as

$$A = \begin{pmatrix} A_1 & & & 0 \\ & A_2 & & \\ 0 & & \ddots & \\ & & & A_r \end{pmatrix} \quad B = \begin{pmatrix} B_1 \\ B_2 \\ \vdots \\ B_r \end{pmatrix} \quad C = (C_1, C_2, \ldots, C_r) \tag{3.50}$$

$$C_i \begin{pmatrix} 1 \\ s \\ \vdots \\ s^{n_i-1} \end{pmatrix} \equiv p_i(s) \pmod{\psi_i(s)} \tag{3.51a}$$

$$[s^{n_i-1} + a_{1i}s^{n_i-2} + \cdots + a_{(n_i-1)i}, \ldots, 1]B_i \equiv \nu_i(s)\mathbf{q}_i^{\mathrm{T}}(s) \pmod{\psi_i(s)} \tag{3.51b}$$

The algorithm is shown in the following example.

Example 3.4 The transfer matrix

$$H(s) = \begin{pmatrix} \dfrac{2s+3}{(s+1)(s+2)} & \dfrac{3s+4}{(s+1)(s+2)} \\ \dfrac{2}{s+1} & \dfrac{1}{s+1} \end{pmatrix}$$

is minimally realized by Kalman's approach as follows. Since $\psi(s) = (s+1)(s+2)$

$$\psi(s)H(s) = \begin{pmatrix} 2s+3 & 3s+4 \\ 2s+4 & 2s+4 \end{pmatrix}$$

Then

$$\begin{pmatrix} 1 & -1 \\ 0 & 1 \end{pmatrix}(\psi(s)H(s)) = \begin{pmatrix} -1 & s \\ 2s+4 & 2s+4 \end{pmatrix}$$

$$\begin{pmatrix} 1 & 0 \\ 2s+4 & 1 \end{pmatrix}\begin{pmatrix} -1 & s \\ 2s+4 & 2s+4 \end{pmatrix} = \begin{pmatrix} -1 & s \\ 0 & 2(s+1)(s+2) \end{pmatrix}$$

$$\begin{pmatrix} -1 & s \\ 0 & 2(s+1)(s+2) \end{pmatrix}\begin{pmatrix} -1 & s \\ 0 & -1 \end{pmatrix} = \begin{pmatrix} 1 & 0 \\ 0 & 2(s+1)(s+2) \end{pmatrix}$$

Giving

$$P^{-1}(s) = \begin{pmatrix} 1 & 0 \\ 2s+4 & 1 \end{pmatrix}\begin{pmatrix} 1 & -1 \\ 0 & 1 \end{pmatrix} = \begin{pmatrix} 1 & -1 \\ 2s+4 & -2s-3 \end{pmatrix}$$

$$Q^{-1}(s) = \begin{pmatrix} -1 & s \\ 0 & 1 \end{pmatrix}$$

Therefore

$$P(s) = \begin{pmatrix} -2s-3 & 1 \\ -2s-4 & 1 \end{pmatrix} \qquad \mathbf{p}_1(s) = \begin{pmatrix} -2s-3 \\ -2s-4 \end{pmatrix}$$

$$Q(s) = \begin{pmatrix} -1 & s \\ 0 & 1 \end{pmatrix} \qquad \mathbf{q}_1^T = [-1 \quad s]$$

Using eqns (3.49), (3.50), and (3.51),

$$\psi(s)H(s) \equiv \mathbf{p}_1(s)\frac{1}{\psi(s)}\mathbf{q}_1^T(s)\psi(s) \pmod{\psi(s)}$$

yields

$$A = \begin{pmatrix} 0 & 1 \\ -2 & -3 \end{pmatrix}, \quad C\begin{pmatrix} 1 \\ s \end{pmatrix} = \begin{pmatrix} -2s-3 \\ -2s-4 \end{pmatrix} \text{giving } C = \begin{pmatrix} -3 & -2 \\ -4 & -2 \end{pmatrix}$$

$$[s+3, 1]B = [-1, s] \text{ giving } B = \begin{pmatrix} 0 & 1 \\ -1 & -3 \end{pmatrix}$$

and the minimal realization is derived.

3.2.3 Mayne's Minimal Realization Algorithm

In Section 3.2.2 one algorithm to minimally realize a given transfer function matrix has been given. But the algorithm requires one to handle polynomial matrices and is not suitable for computer computation. This section gives Mayne's algorithm which can easily be achieved computationally. The algorithm consists of two steps. In the first step a controllable (observable) system is realized for the given transfer function and in the second step, its observable (controllable) subsystem is computed to give a minimal realization. The procedure is completely dependent on Theorem 2.6 and (3.35) or Theorem 2.7 and (2.40).

(1) Realization of a controllable form and its minimal realization by picking its observable subsystem

When the transfer function matrix $H(s)$ is given by (3.38), let us consider its ith column

$$h_i(s) = \begin{pmatrix} \dfrac{p_{1i}(s)}{q_{1i}(s)} \\ \dfrac{p_{2i}(s)}{q_{2i}(s)} \\ \vdots \\ \dfrac{p_{pi}(s)}{q_{pi}(s)} \end{pmatrix}$$

and let the least common multiple of $q_{1i}(s), \ldots, q_{pi}(s)$ be $d_i(s)$, then $h_i(s)$ can be written as

$$h_i(s) = \frac{1}{d_i(s)} \begin{pmatrix} n_{1i}(s) \\ n_{2i}(s) \\ \vdots \\ n_{pi}(s) \end{pmatrix}$$

where

$$\begin{aligned} n_{ji}(s) &= \beta_{ji(n_i-1)} s^{n_i-1} + \beta_{ji(n_i-2)} s^{n_i-2} + \cdots + \beta_{ji0} \\ d_i(s) &= s^{n_i} + \alpha_{i(n_i-1)} s^{n_i-1} + \cdots + \alpha_{i0} \end{aligned}$$

From (3.19), $h_i(s)$ can be realized in the controllable canonical form as

$$A_i = \begin{pmatrix} 0 & 1 & & 0 \\ \vdots & & \ddots & \\ 0 & 0 & & 1 \\ -\alpha_{i0} & & \cdots & -\alpha_{i(n_i-1)} \end{pmatrix} \qquad \mathbf{b}_i = \begin{pmatrix} 0 \\ \vdots \\ 0 \\ 1 \end{pmatrix}$$

$$C_i = \begin{pmatrix} \beta_{1i0} & \beta_{1i1} & \cdots & \beta_{1i(n_i-1)} \\ \vdots & & & \vdots \\ \beta_{pi0} & \beta_{pi1} & \cdots & \beta_{pi(n_i-1)} \end{pmatrix} \tag{3.52}$$

Thereafter $H(s)$ is realized as follows. Writing

$$A = \begin{pmatrix} A_1 & & & \\ & A_2 & & 0 \\ & & \ddots & \\ 0 & & & A_m \end{pmatrix}, \quad B = \begin{pmatrix} \mathbf{b}_1 & & & \\ & \mathbf{b}_2 & & 0 \\ & & \ddots & \\ 0 & & & \mathbf{b}_m \end{pmatrix}$$

$$C^{\mathrm{T}} = \begin{pmatrix} C_1 \\ C_2 \\ \vdots \\ C_m \end{pmatrix}. \tag{3.53}$$

The system (3.53) is controllable, but it may not be observable. So its observable subsystem as given in Theorem 2.7 should be taken. Since the observability matrix of (3.53) is

$$\mathcal{O}^{\mathrm{T}} = \begin{pmatrix} C \\ CA \\ CA^2 \\ \vdots \\ CA^{n-1} \end{pmatrix}$$

and if rank $\mathcal{O}^{\mathrm{T}} = n_0 < n$, then n_0 linearly independent row vectors $\mathbf{w}_1^{\mathrm{T}}, \mathbf{w}_2^{\mathrm{T}}, \ldots, \mathbf{w}_n^{\mathrm{T}}$ can be taken from $\mathcal{O}^{\mathrm{T}}$. Let these vectors form the matrix, S, where

$$S = \begin{pmatrix} \mathbf{w}_1^{\mathrm{T}} \\ \mathbf{w}_2^{\mathrm{T}} \\ \vdots \\ \mathbf{w}_{n_0}^{\mathrm{T}} \end{pmatrix}.$$

and find the matrix U satisfying

$$SU = I_{n_0} \tag{3.54}$$

then

$$A_0 = SAU,\ B_0 = SB,\ C_0 = CU \tag{3.55}$$

gives the observable and controllable system with the given transfer function.

Remark The procedure is to find $(\bar{A}_{11}, \bar{B}_1, \bar{C}_0)$ in Theorem 2.7 and U consisting of column vectors which are not included in $N(\mathcal{O}^{\mathrm{T}})$. To find such a U one chooses S' to make $\binom{S}{S'}$ non-singular and $\binom{S}{S'}^{-1} = [U, U']$ gives U.

Example 3.5 Find a minimal realization of

$$H(s) = \begin{pmatrix} \dfrac{4s+6}{(s+1)(s+2)} & \dfrac{2s+3}{(s+1)(s+2)} \\ \dfrac{-2}{(s+1)(s+2)} & \dfrac{-1}{(s+1)(s+2)} \end{pmatrix}$$

From (3.52) the controllable realization for $\mathbf{h}_1(s)$ is

$$A_1 = \begin{pmatrix} 0 & 1 \\ -2 & -3 \end{pmatrix} \quad \mathbf{b}_1 = \begin{pmatrix} 0 \\ 1 \end{pmatrix} \quad C_1 = \begin{pmatrix} 6 & 4 \\ -2 & 0 \end{pmatrix}$$

and that for $\mathbf{h}_2(s)$ is

$$A_2 = \begin{pmatrix} 0 & 1 \\ -2 & -3 \end{pmatrix} \quad \mathbf{b}_2 = \begin{pmatrix} 0 \\ 1 \end{pmatrix} \quad C_2 = \begin{pmatrix} 3 & 2 \\ -1 & 0 \end{pmatrix}$$

They give a controllable realization

$$A = \begin{pmatrix} -0 & 1 & 0 & 0 \\ -2 & -3 & 0 & 0 \\ 0 & 0 & 0 & 1 \\ 0 & 0 & -2 & -3 \end{pmatrix} \quad B = \begin{pmatrix} 0 & 0 \\ 1 & 0 \\ 0 & 0 \\ 0 & 1 \end{pmatrix} \quad C = \begin{pmatrix} 6 & 4 & 3 & 2 \\ -2 & 0 & -1 & 0 \end{pmatrix}$$

The observability matrix of (A, B, C) is

$$\mathcal{O}^{\mathrm{T}} = \begin{pmatrix} 6 & 4 & 3 & 2 \\ -2 & 0 & -1 & 0 \\ -8 & -6 & -4 & -3 \\ 0 & -2 & 0 & -1 \\ \vdots & \vdots & \vdots & \vdots \end{pmatrix}$$

CA is linearly dependent on C, so CA^2 and CA^3 are also linearly dependent on C, which gives rank $\mathcal{O}^{\mathrm{T}} = 2$, so that $\mathbf{w}_1^{\mathrm{T}} = [6, 4, 3, 2]$, and $\mathbf{w}_2^{\mathrm{T}} = [-2, 0, -1, 0]$. S is given by

$$S = \begin{pmatrix} \mathbf{w}_1^{\mathrm{T}} \\ \mathbf{w}_2^{\mathrm{T}} \end{pmatrix} = \begin{pmatrix} 6 & 4 & 3 & 2 \\ -2 & 0 & -1 & 0 \end{pmatrix}$$

and $U = S^{-1}$ by

$$U = \frac{1}{8} \begin{pmatrix} 0 & -4 \\ 2 & 6 \\ 0 & 0 \\ 0 & 0 \end{pmatrix}.$$

Therefore a minimal realization

$$A_0 = SAU = \frac{1}{2}\begin{pmatrix} -3 & -1 \\ -1 & -3 \end{pmatrix}, \; B_0 = SB = \begin{pmatrix} 4 & 2 \\ 0 & 0 \end{pmatrix}, \; C_0 = CU = \begin{pmatrix} 1 & 0 \\ 0 & 1 \end{pmatrix}$$

is derived.

The dual of the above procedure can also be considered.

(2) Realization of an observable form and its minimal realization by picking its controllable subsystem

Let the ith row of $H(s)$ be

$$\mathbf{h}_i^{\mathsf{T}} = \begin{pmatrix} \frac{p_{i1}(s)}{q_{i1}(s)} & \frac{p_{i2}(s)}{q_{i2}(s)} & \dots & \frac{p_{im}(s)}{q_{im}(s)} \end{pmatrix}$$

and let $d_i(s)$ be the least common multiple of the denominators $q_{ij}(s)$. Then it can be rewritten as

$$\mathbf{h}_i^{\mathsf{T}} = \frac{1}{d_i(s)}\,[n_{i1}(s), n_{i2}(s), \dots, n_{im}(s)]$$

where

$$n_{ij}(s) = \beta_{ij(n_i-1)}s^{n_i-1} + \beta_{ij(n_i-2)}s^{n_i-2} + \cdots + \beta_{ij0}$$
$$d_i(s) = s^{n_i} + \alpha_{i(n_i-1)}s^{n_i-1} + \alpha_{i(n_i-2)}s^{n_i-2} + \cdots + \alpha_{i0}$$

From the transpose of (3.27) the observable realization of $\mathbf{h}_i^{\mathsf{T}}(s)$ is given by

$$A_i = \begin{pmatrix} 0 & \dots & 0 & -\alpha_{i0} \\ 1 & \ddots & 0 & \\ 0 & & 1 & -\alpha_{in_{i-1}} \end{pmatrix} B_i = \begin{pmatrix} \beta_{i10} & \dots & \beta_{im0} \\ \vdots & & \\ \beta_{i1n_{i-1}} & \dots & \beta_{imn_{i-1}} \end{pmatrix}$$
$$\mathbf{c}_i = [0, 0, \dots, 1]. \tag{3.56}$$

$H(s)$ is then realized by

$$A = \begin{pmatrix} A_1 & & & \\ & A_2 & & 0 \\ & & \ddots & \\ 0 & & & A_p \end{pmatrix} \quad B = \begin{pmatrix} B_1 \\ B_2 \\ \vdots \\ B_p \end{pmatrix} \quad C = \begin{pmatrix} \mathbf{c}_1^{\mathsf{T}} & \mathbf{0}^{\mathsf{T}} & \dots & \mathbf{0}^{\mathsf{T}} \\ \mathbf{0}^{\mathsf{T}} & \mathbf{c}_2^{\mathsf{T}} & \dots & \mathbf{0}^{\mathsf{T}} \\ \mathbf{0}^{\mathsf{T}} & & \dots & \mathbf{c}_p^{\mathsf{T}} \end{pmatrix} \tag{3.57}$$

The controllable subsystem of (A, B, C) is given as follows. Let the rank of the controllability matrix

$$\mathscr{C} = [B, AB, \dots, A^{n-1}B]$$

be $n_c < n$. Then choose n_c linearly independent column vectors $\mathbf{v}_1, \dots, \mathbf{v}_{n_c}$ from $B, AB, \dots, A^{n-1}B$. Let U be a matrix formed from these vectors where

$$U = [\mathbf{v}_1, \mathbf{v}_2, \dots, \mathbf{v}_{n_c}]$$

and find S such that

$$SU = I_{n_c}$$

then

$$A_c = SAU, \; B_c = SB, \; C_c = CU \qquad (3.58)$$

gives a minimal realization.

APPENDIX

The Proof of Theorem 3.2

To prove (3.31),

$$B_c \hat{B}_m^{-1} \delta_c(s) = (sI - A_c)S(s) \qquad (3.31)'$$

will be proved. The left of (3.31)′ is

$$(sI - A_c)S(s) = \begin{pmatrix} s & -1 & & 0 & & & \\ & & \ddots & & & & \\ 0 & & & -1 & & & \\ \alpha_{10} & & \dots & s + \alpha_{1(\sigma_1 - 1)} & \dots & & \alpha_{1(n-1)} \\ 0 & & & & & & \\ \vdots & & & & \ddots & & \vdots \\ 0 & & & & & s & -1 \\ \alpha_{m0} & & & & \dots & & s + \alpha_{m(n-1)} \end{pmatrix} \begin{pmatrix} 1 & \dots & 0 \\ s & & \vdots \\ \vdots & & \\ s^{\sigma_1 - 1} & & \\ 0 & & \\ \vdots & & \\ 0 & \dots & s^{\sigma_m - 1} \end{pmatrix}$$

$$= \begin{pmatrix} & 0^{\mathrm{T}} & \\ s^{\sigma_1} + \sum_{i=0}^{\sigma_1 - 1} \alpha_{1i} s^i & \sum_{i=0}^{\sigma_2 - 1} \alpha_{1(i+\sigma_1)} s^i & \dots \\ & 0^{\mathrm{T}} & \\ & \vdots & \\ & 0^{\mathrm{T}} & \\ \sum_{i=0}^{\sigma_1 - 1} \alpha_{mi} s^i & \dots & \end{pmatrix}$$

Therefore (3.31)′ exists. By dividing (3.31)′ by $(sI - A_c)$ from the left and by $\delta_c(s)$ from the right, and by multiplying by C_c from the left gives (3.31).

The Proof of Theorem 3.4

Let both (A_1, B_1, C_1) and (A_2, B_2, C_2) be minimal realizations for $H(s)$, then

$$H_i = C_1 A_1^i B_1 = C_2 A_2^i B_2 \qquad i = 0, 1, 2, \dots \qquad (3.59)$$

Let

$$\mathscr{C}_1 = [B_1, A_1B_1, \ldots, A_1^{n-1}B_1]$$

$$\mathscr{C}_2 = [B_2, A_2B_2, \ldots, A_2^{n-1}B_2]$$

then (3.59) gives

$$C_1\mathscr{C}_1 = C_2\mathscr{C}_2 \tag{3.60}$$

and from the fact that they are minimal realizations, that is, controllable and observable.

$$\text{rank } \mathscr{C}_1 = \text{rank } \mathscr{C}_2 = n$$

Let

$$T_1 = \mathscr{C}_1\mathscr{C}_2^{\mathrm{T}}(\mathscr{C}_2\mathscr{C}_2^{\mathrm{T}})^{-1}$$

then

$$C_1T_1 = C_2 \tag{3.61}$$

From (3.59),

$$\mathscr{O}_1^{\mathrm{T}}B_1 = \mathscr{O}_2^{\mathrm{T}}B_2 \tag{3.62}$$

where

$$\mathscr{O}_1 = [C_1^{\mathrm{T}}, A_1^{\mathrm{T}}C_1^{\mathrm{T}}, (A_1^{\mathrm{T}})^2C_1{}^{\mathrm{T}}, \ldots, (A_1^{\mathrm{T}})^{n-1}C_1^{\mathrm{T}}]$$

$$\mathscr{O}_2 = [C_2^{\mathrm{T}}, A_2^{\mathrm{T}}C_2^{\mathrm{T}}, (A_2^{\mathrm{T}})^2C_2^{\mathrm{T}}, \ldots, (A_2^{\mathrm{T}})^{n-1}C_2^{\mathrm{T}}].$$

(3.62) gives

$$T_2B_1 = B_2 \tag{3.63}$$

where

$$T_2 = (\mathscr{O}_2\mathscr{O}_2^{\mathrm{T}})^{-1}\mathscr{O}_2\mathscr{O}_1^{\mathrm{T}}$$

and (3.62) yields

$$\mathscr{O}_1^{\mathrm{T}}\mathscr{C}_1 = \mathscr{O}_2^{\mathrm{T}}\mathscr{C}_2$$

$$\mathscr{O}_1^{\mathrm{T}}A_1\mathscr{C}_1 = \mathscr{O}_2^{\mathrm{T}}A_2\mathscr{C}_2 \tag{3.64}$$

From the above equations, the following equation is derived.

$$T_2AT_1 = A_2. \tag{3.65}$$

Now (3.64) gives

$$\begin{aligned} T_2 T_1 &= (\mathcal{O}_2 \mathcal{O}_2^{\mathrm{T}})^{-1} \mathcal{O}_2 \mathcal{O}_1^{\mathrm{T}} \mathcal{C}_1 \mathcal{C}_2^{\mathrm{T}} (\mathcal{C}_2 \mathcal{C}_2^{\mathrm{T}})^{-1} \\ &= (\mathcal{O}_2 \mathcal{O}_2^{\mathrm{T}})^{-1} \mathcal{O}_2 \mathcal{O}_2^{\mathrm{T}} \mathcal{C}_2 \mathcal{C}_2^{\mathrm{T}} (\mathcal{C}_2 \mathcal{C}_2^{\mathrm{T}})^{-1} \\ &= I \end{aligned}$$

therefore

$$T_2 = T_1^{-1}$$

Thus (3.61), (3.63), (3,65) show that (A_1, B_1, C_1) and (A_2, B_2, C_2) are equivalent.

PROBLEMS

P3.1 Find Luenberger's second Controllable Canonical Form for the following systems.

(1–1)

$$(A, B, \mathbf{c}^{\mathrm{T}}) = \left[\begin{pmatrix} 0 & 1 & 2 \\ 1 & 3 & 4 \\ 0 & 5 & 6 \end{pmatrix}, \begin{pmatrix} 1 & 0 \\ 0 & 0 \\ 0 & 1 \end{pmatrix}, [(0 \quad 1 \quad 0) \right]$$

(1–2)

$$(A, \mathbf{b}, \mathbf{c}^{\mathrm{T}}) = \left[\begin{pmatrix} -1 & 1 \\ 0 & -2 \end{pmatrix}, \begin{pmatrix} 0 \\ 1 \end{pmatrix}, [(1 \quad 0)] \right]$$

(1–3)

$$(A, B, \mathbf{c}^{\mathrm{T}}) = \left[\begin{pmatrix} 0 & 0 & 1 \\ 0.5 & 0 & 0 \\ 0 & 2 & 0 \end{pmatrix}, \begin{pmatrix} 0.5 & -0.5 \\ 0.25 & 0.25 \\ -0.5 & 0.5 \end{pmatrix}, [1 \quad 2 \quad 0] \right]$$

(1–4)

$$(A, B, C) = \left[\begin{pmatrix} -3 & -2 & 4 & 3 \\ -2 & -3 & 4 & 4 \\ 0 & 0 & 0 & 1 \\ -6 & -5 & 9 & 6 \end{pmatrix}, \begin{pmatrix} 0 & 1 \\ 1 & -1 \\ 0 & 0 \\ 1 & 0 \end{pmatrix}, \begin{pmatrix} -1 & 0 & 2 & 1 \\ 2 & 2 & -2 & -1 \end{pmatrix} \right]$$

P3.2 Find the transfer functions of the systems in Problem P3.1.

P3.3 Obtain minimal realizations for the following transfer matrices.

(3–1)

$$H(s) = \begin{pmatrix} \dfrac{2s+3}{s^2+3s+2} & \dfrac{1}{s^2+3s+2} \\[2ex] \dfrac{3}{s+1} & \dfrac{3}{s+1} \end{pmatrix}$$

(3–2)

$$H(s)=\begin{pmatrix}\dfrac{s+2}{s^2+2s+1} & \dfrac{-1}{s^2+3s+2}\\[2ex] \dfrac{-s+1}{s^2+2s+1} & \dfrac{1}{s^2+3s+2}\end{pmatrix}$$

(3–3)

$$H(s)=\begin{pmatrix}\dfrac{2s+3}{s^2+3s+2} & \dfrac{1}{s+2} & \dfrac{1}{s+1}\\[2ex] \dfrac{1}{s^2+3s+2} & \dfrac{-1}{s^2+5s+6} & \dfrac{2s+4}{s^2+4s+3}\end{pmatrix}$$

(3–4) $$H(s)=\begin{pmatrix}\dfrac{2s^2+3s+2}{s^2+2s+1} & \dfrac{s+3}{s+1}\end{pmatrix}$$

P3.4 If the matrix A is in the companion form

$$A=\begin{pmatrix}0 & & & \\ \vdots & & I_{n-1} & \\ 0 & & & \\ -\alpha_0 & -\alpha_1 & \cdots & -\alpha_{n-1}\end{pmatrix}$$

and $\psi(s)=s^n+\alpha_{n-1}s^{n-1}+\cdots+\alpha_0$, then prove the following relation.

$$\psi(s)[sI-A]^{-1}\equiv\begin{pmatrix}1\\ s\\ s^2\\ \vdots\\ s^{n-1}\end{pmatrix}[s^{n-1}+\alpha_{n-1}s^{n-2}+\cdots+\alpha_1,\ s^{n-2}+\alpha_{n-1}s^{n-3}$$

$$+\cdots+\alpha_2,\cdots,1]\,(\mathrm{mod}\ \psi(s)) \qquad (3.66)$$

where the equality $\equiv$ (mod $\psi(s)$) means that the residues of both sides divided by $\psi(s)$ are equal.

P3.5 For Markov parameters $H_1, H_2, \ldots$, if the Hankel matrix of (3.43) is assumed to satisfy

$$\mathrm{rank}\ \mathcal{H}_r=\mathrm{rank}\ \mathcal{H}_{r+1}=\cdots=\mathrm{rank}\ \mathcal{H}_n=n$$

for a certain r, then $\mathcal{H}_n$ can be expressed as the product of an $np\times n$ matrix P and an $n\times nm$ matrix as

$$\mathcal{H}_n=PQ$$

where rank $P=$ rank $Q=n$.

Let τ be defined as

$$\tau\mathscr{H}_n = \begin{pmatrix} H_1 & H_2 & \dots & H_n \\ \vdots & & & \\ H_n & H_{n+1} & \dots & H_{2n+1} \end{pmatrix}$$

then show that there exists an $n \times nm$ matrix $Q^{\#}$ such that

$\tau\mathscr{H}_n = PQ^{\#}$

and $H_1, H_2, \dots$ are minimally realized by

$$A = Q^{\#}Q^{\mathrm{T}}(QQ^{\mathrm{T}})^{-1}$$

$$B = Q\begin{pmatrix} I_m \\ \vdots \\ 0 \end{pmatrix} \qquad C = (I_p \quad 0 \quad \dots)P$$

P3.6 A system has

$$(A, B, C, D) = \left[\begin{pmatrix} 1 & 2 & 0 \\ 4 & -1 & 0 \\ 0 & 0 & 1 \end{pmatrix}, \begin{pmatrix} 1 \\ 0 \\ 1 \end{pmatrix}, \begin{pmatrix} 0 & 1 & -1 \\ 0 & 0 & 1 \end{pmatrix}, \begin{pmatrix} 0 \\ 1 \end{pmatrix}\right]$$

Determine the input–output transfer function matrix. Is the system completely characterized by the transfer matrix? Confirm your answer by showing from the above matrices that the system is controllable and observable.

P3.7 A multivariable system has a transfer matrix

$$H(s) = \begin{pmatrix} \dfrac{2}{(s+2)} & \dfrac{4s-1}{(s+2)(s+3)} \\ \dfrac{3}{(s+3)} & \dfrac{5}{(s+3)} \end{pmatrix}$$

Find its minimal state space representation.

4
STATE FEEDBACK AND DECOUPLING

4.1 STATE FEEDBACK

4.1.1 State Feedback and the Controllable Subspace

Consider the following n-dimensional linear system with m-inputs and p-outputs

$$\dot{\mathbf{x}} = A\mathbf{x} + B\mathbf{u} \tag{4.1a}$$

$$\mathbf{y} = C\mathbf{x} \tag{4.1b}$$

For this system the effect of implementing a control law consisting of state feedback $F\mathbf{x}$ and feedforward $G\mathbf{v}$ is considered, that is

$$\mathbf{u} = F\mathbf{x} + G\mathbf{v} \tag{4.2}$$

We discuss how the structure of the system is changed by this control. Substituting (4.2) into (4.1) we have

$$\dot{\mathbf{x}} = (A + BF)\mathbf{x} + BG\mathbf{v} \tag{4.3a}$$

$$\mathbf{y} = C\mathbf{x}. \tag{4.3b}$$

Thus by the control of (4.2) (A, B, C) is changed to $(A + BF, BG, C)$. The structure of the closed loop system is depicted in Figure 4.1.

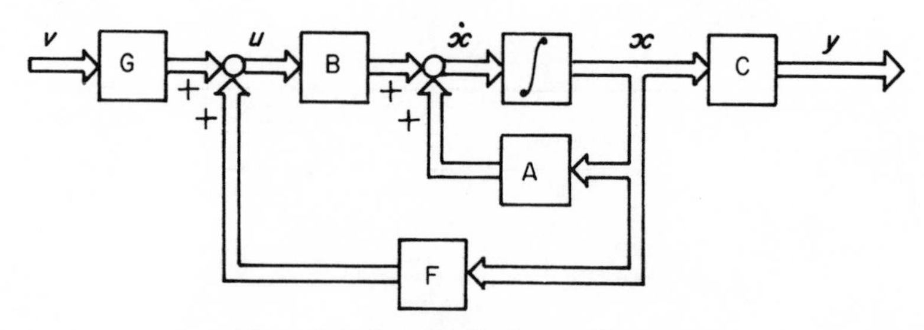

Figure 4.1 State feedback control system

Theorem 4.1

The controllable subspace of (A, B, C) is the same as that of $(A + BF, BG, C)$ with any F and non-singular G.

Before the theorem is proved, the following example may be useful to illustrate this property.

Example 4.1 The controllable subspace of the SISO system

$$\dot{\mathbf{x}} = \begin{pmatrix} 0 & -2 \\ 1 & -3 \end{pmatrix}\mathbf{x} + \begin{pmatrix} 1 \\ 1 \end{pmatrix} u$$

is given by

$$\mathscr{X}_c = \mathscr{R}\begin{pmatrix} 1 & -2 \\ 1 & -2 \end{pmatrix} = \left\{\begin{pmatrix} 1 \\ 1 \end{pmatrix}\right\}$$

When the control is given by

$$u = (f_1, f_2)\mathbf{x} + gv(g \neq 0)$$

the system becomes

$$\dot{\mathbf{x}} = \begin{pmatrix} f_1, & f_2 - 2 \\ 1 + f_1, & f_2 - 3 \end{pmatrix}\mathbf{x} + g\begin{pmatrix} 1 \\ 1 \end{pmatrix} v$$

The controllable subspace

$$\mathscr{X}_c = \mathscr{R}\begin{pmatrix} 1, f_1 + f_2 - 2 \\ 1, f_1 + f_2 - 2 \end{pmatrix} = \left\{\begin{pmatrix} 1 \\ 1 \end{pmatrix}\right\}$$

which we see is not changed by the state feedback.

Proof of Theorem 4.1 Let the controllability matrix of (A, B, C) be

$$\mathscr{C}_1 = (B, AB, \ldots, A^{n-1}B)$$

and that of $(A + BF, BG, C)$ be $\mathscr{C}_2$. Then the controllable subspaces are given by $\mathscr{R}(\mathscr{C}_1)$ and $\mathscr{R}(\mathscr{C}_2)$ respectively. The non-zero vector $\mathbf{x}_1$ which is orthogonal to $\mathscr{R}(\mathscr{C}_1)$ satisfies $\mathbf{x}_1^{\mathrm{T}}\mathscr{C}_1 = 0$, which yields $\mathbf{x}_1^{\mathrm{T}}A^iB = 0$ for $i = 0, 1, 2, \ldots, n-1$. Now

$$\mathbf{x}_1^{\mathrm{T}}BG = \mathbf{0}^{\mathrm{T}}, \text{so that } \mathbf{x}_1^{\mathrm{T}}(A + BF)BG = \mathbf{x}_1^{\mathrm{T}}ABG = \mathbf{0}^{\mathrm{T}}$$

and

$$\begin{aligned} \mathbf{x}_1^{\mathrm{T}}(A + BF)^iBG &= \mathbf{x}_1^{\mathrm{T}}(A + BF)(A + BF)^{i-1}BG \\ &= \mathbf{x}_1^{\mathrm{T}}A(A + BF)^{i-1}BG = \cdots = \mathbf{x}_1^{\mathrm{T}}A^iBG = \mathbf{0}^{\mathrm{T}} \end{aligned}$$

which yields

$$\mathbf{x}_1^{\mathrm{T}}\mathscr{C}_2 = \mathbf{0}^{\mathrm{T}}$$

and $\mathcal{R}(\mathcal{C}_2)^{\perp} \supset \mathcal{R}(\mathcal{C}_1)^{\perp}$ is derived which is equivalent to

$$\mathcal{R}(\mathcal{C}_1) \supset \mathcal{R}(\mathcal{C}_2)$$

The converse statement is proved as follows. $\mathbf{x}_2^{\mathrm{T}} \mathcal{C}_2 = \mathbf{0}$ means $\mathbf{x}_2^{\mathrm{T}}(A + BF)^i BG = \mathbf{0}$ for $i = 0, 1, 2, \ldots, n-1$. Since G is non-singular

$$\begin{aligned}
\mathbf{0}^{\mathrm{T}} &= \mathbf{x}_2^{\mathrm{T}} B, \mathbf{0}^{\mathrm{T}} = \mathbf{x}_2^{\mathrm{T}} AB, \ldots, \\
\mathbf{0}^{\mathrm{T}} &= \mathbf{x}_2^{\mathrm{T}}(A + BF)^i BG = \mathbf{x}_2^{\mathrm{T}}(A + BF)(A + BF)^{i-1} BG \\
&= \mathbf{x}_2^{\mathrm{T}} A (A + BF)^{i-1} BG = \cdots = \mathbf{x}_2^{\mathrm{T}} A^i BG
\end{aligned}$$

Therefore

$$\mathcal{R}(\mathcal{C}_1) \subset \mathcal{R}(\mathcal{C}_2)$$

and

$$\mathcal{R}(\mathcal{C}_1) = \mathcal{R}(\mathcal{C}_2)$$

4.1.2 Pole Assignment by State Feedback

Theorem 4.1 shows the fact that the controllable subspace does not vary with state feedback, which indicates that the controllability of the system $(A + BF, B, C)$ is the same as that of the system (A, B, C). However the poles, that is the mode characteristics of the state space, can be changed by state feedback. This is illustrated by the following example.

Example 4.2 For a controllable single variable system

$$\dot{\mathbf{x}} = A\mathbf{x} + \mathbf{b}u$$

where the characteristic equation of A is

$$\begin{aligned}
\phi(s) &= \det(Is - A) \\
&= \prod_{i=1}^{n} (s - s_i) \\
&= s^n + \alpha_{n-1}s^{n-1} + \cdots + \alpha_0
\end{aligned}$$

We determine the control law

$$u = \mathbf{f}^{\mathrm{T}}\mathbf{x} + v$$

so that the characteristic equation of the closed loop system on substituting the control law, that is

$$\dot{\mathbf{x}} = (A + \mathbf{b}\mathbf{f}^{\mathrm{T}})\mathbf{x} + \mathbf{b}v$$

has a predetermined characteristic equation of the form

$$\begin{aligned}
\phi_f(s) &= \det(sI - A - \mathbf{b}\mathbf{f}^{\mathrm{T}}) \\
&= s^n + \gamma_{n-1}s^{n-1} + \cdots + \gamma_0
\end{aligned}$$

The required $\mathbf{f}^{\mathsf{T}}$ can be determined as follows (Gopinath). The relationships

$$\det\begin{pmatrix} sI - A & \mathbf{b} \\ \mathbf{f}^{\mathsf{T}} & 1 \end{pmatrix} = \det(sI - A - \mathbf{b}\mathbf{f}^{\mathsf{T}}) = \det\begin{pmatrix} sI - A & \mathbf{b} \\ \mathbf{0}^{\mathsf{T}} & 1 - \mathbf{f}^{\mathsf{T}}(sI - A)^{-1}\mathbf{b} \end{pmatrix}$$

yield

$$\det(sI - A - \mathbf{b}\mathbf{f}^{\mathsf{T}}) = \det(sI - A)(1 - \mathbf{f}^{\mathsf{T}}(sI - A)^{-1}\mathbf{b})$$

so that

$$\begin{aligned} s^n + \gamma_{n-1}s^{n-1} + \cdots + \gamma_0 \\ = s^n + \alpha_{n-1}s^{n-1} + \cdots + \alpha_0 - \mathbf{f}^{\mathsf{T}}(\Gamma_{n-1}s^{n-1} + \Gamma_{n-2}s^{n-2} + \cdots + \Gamma_0)\mathbf{b} \end{aligned} \tag{4.4}$$

where

$$(sI - A)^{-1} = \frac{1}{\phi(s)}(\Gamma_{n-1}s^{n-1} + \cdots + \Gamma_0)$$

Comparing coefficients of powers of s in (4.4) gives

$$\mathbf{f}^{\mathsf{T}}[\Gamma_{n-1}\mathbf{b}, \Gamma_{n-2}\mathbf{b}, \ldots, \Gamma_0\mathbf{b}] = [\alpha_{n-1} - \gamma_{n-1}, \ldots, \alpha_0 - \gamma_0].$$

The Γ_i are given by Faddeev's algorithm, (1.26), that is

$$\Gamma_{n-1} = I, \Gamma_{n-2} = A + \alpha_{n-1}I, \ \Gamma_{n-3} = A^2 + \alpha_{n-1}A + \alpha_{n-2}I, \ldots$$

so that $[\Gamma_{n-1}\mathbf{b}, \Gamma_{n-2}\mathbf{b}, \ldots, \Gamma_0\mathbf{b}]$ can be written

$$[\Gamma_{n-1}\mathbf{b}, \ldots, \Gamma_0\mathbf{b}] = \mathscr{C}\begin{pmatrix} 1 & \alpha_{n-1} & & \alpha_1 \\ 0 & 1 & & \alpha_2 \\ \vdots & 0 & \ddots & \vdots \\ & \vdots & & \alpha_{n-1} \\ 0 & 0 & & 1 \end{pmatrix}$$

where

$$\mathscr{C} = [\mathbf{b}, A\mathbf{b}, \ldots, A^{n-1}\mathbf{b}]$$

Since the system is controllable, $\mathscr{C}$ is non-singular and we have the required result

$$\mathbf{f}^{\mathsf{T}} = (\alpha_{n-1} - \gamma_{n-1}, \ldots, \alpha_0 - \gamma_0)\begin{pmatrix} 1 & \alpha_{n-1} & & \alpha_1 \\ 0 & 1 & & \vdots \\ \vdots & \vdots & \ddots & \\ 0 & 0 & & 1 \end{pmatrix}^{-1}\mathscr{C}^{-1}. \tag{4.5}$$

The algorithm (4.5) includes the parameters of the characteristic equation of A which therefore have to be evaluated. Ackermann proposed another algorithm of pole allocation for single input systems where $\mathbf{f}^{\mathsf{T}}$ is given by

$$\mathbf{f}^{\mathsf{T}} = -[0, 0, \ldots, 0, 1]\mathscr{C}^{-1}\phi_f(\mathrm{A}). \tag{4.6}$$

The proof is as follows. We write the relationship between the system with state feedback and its controllable canonical form, that is

$$T^{-1}(A + \mathbf{b}\mathbf{f}^{\mathsf{T}}) = \begin{pmatrix} 0 & 1 & & \\ & & & 1 \\ -\gamma_0 & \cdots & & \gamma_{n-1} \end{pmatrix}T^{-1}.$$

where T^{-1} is given by (3.5) as

$$T^{-1} = \begin{pmatrix} \mathbf{l}_n^{\mathrm{T}} \\ \mathbf{l}_n^{\mathrm{T}} A \\ \vdots \\ \mathbf{l}_n^{\mathrm{T}} A^{n-1} \end{pmatrix} \text{with } \mathbf{l}_n^{\mathrm{T}} = [0, \ldots, 0, 1]\mathscr{C}^{-1}$$

On premultiplying both sides of the above relationship by $(0, \ldots, 0, 1)$ we obtain (4.6). Equations (4.5) and (4.6) indicate that for any given γ_i, the corresponding state feedback $\mathbf{f}^{\mathrm{T}}$ can be determined if and only if the system is controllable. This result can be extended to multivariable systems as originally shown by Wonham.

Theorem 4.2

For a linear multivariable system (A, B, C) a state feedback F can be found such that the characteristic equation of the feedback system has arbitrary real coefficients if and only if (A, B) is controllable.

Proof Since (A, B, C) has the same characteristic equation as its equivalent system $(\bar{A}, \bar{B}, \bar{C})$

$$\det(sI - \bar{A}) = \det T^{-1} \det(sI - A) \det T.$$

Also the feedback law F of the system is related to that of the equivalent system $\bar{F}$ by $\bar{F}T = F$, and the following relationship exists

$$\det(sI - \bar{A} - \bar{B}\bar{F}) = \det T^{-1} \det(sI - A - BF) \det T$$

This equation shows that the pole assignability of (A, B, C) is the same as that of $(\bar{A}, \bar{B}, \bar{C})$, and if (A, B) is not controllable, then from Theorem 2.6, it has an equivalent system of the form

$$\bar{A} = \begin{bmatrix} \bar{A}_{\mathrm{c}} & \bar{A}_{12} \\ 0 & \bar{A}_{22} \end{bmatrix} \quad \bar{B} = \begin{bmatrix} \bar{B}_{\mathrm{c}} \\ 0 \end{bmatrix} \quad \bar{F} = [\bar{F}_{\mathrm{c}}, \bar{F}_2] \tag{4.7}$$

The characteristic determinant of this system is given by

$$\det(sI - \bar{A} - \bar{B}\bar{F}) = \det\begin{pmatrix} sI - \bar{A}_{\mathrm{c}} - \bar{B}_{\mathrm{c}}\bar{F}_{\mathrm{c}} & -\bar{A}_{12} - \bar{B}_{\mathrm{c}}\bar{F}_2 \\ 0 & sI - \bar{A}_{22} \end{pmatrix}$$

$$= \det(sI - \bar{A}_{\mathrm{c}} - \bar{B}_{\mathrm{c}}\bar{F}_{\mathrm{c}})\det(sI - \bar{A}_{22})$$

which shows that all the characteristic roots of A cannot be changed by state feedback. Controllability of (A, B) is thus a necessary condition for all the poles to be assigned arbitrarily. Sufficiency can be proved by determining the feedback law F for a controllable system (A, B) so that the resulting closed loop system has a preassigned characteristic equation. Let the controllability indices of the system be $\{\sigma_i\}$, then one of the equivalent systems can be represented by the controllable canonical form

$$\bar{A} = \begin{pmatrix} \mathbf{E}_1 \\ \mathbf{a}_1^T \\ \mathbf{E}_2 \\ \mathbf{a}_2^T \\ \vdots \\ \mathbf{E}_m \\ \mathbf{a}_m^T \end{pmatrix} \qquad \bar{B} = \begin{pmatrix} 0 \\ \mathbf{b}_1^T \\ 0 \\ \mathbf{b}_2^T \\ \vdots \\ 0 \\ \mathbf{b}_m^T \end{pmatrix}$$

$$\left.\begin{aligned} &\mathbf{E}_i = [\underbrace{0, \ldots, 0}_{\sum_{j=1}^{i-1} \sigma_j + 1}, 0, I_{\sigma_i - 1} 0, \ldots, \underbrace{0}_{\sigma_m}]\}\sigma_i - 1 \\ &\mathbf{a}_i^T = [\alpha_{i0}, \alpha_{i1}, \ldots, \alpha_{im-1}] \\ &\mathbf{b}_i^T = [0, \ldots, 0, 1, \beta_{ii+1}, \ldots, \beta_{im}] \end{aligned}\right\} (i = 1, \ldots, m)$$

Let

$$\hat{B}_m = \begin{pmatrix} \mathbf{b}_1^T \\ \vdots \\ \mathbf{b}_m^T \end{pmatrix}$$

and define

$$\bar{F} = \hat{B}_m^{-1} \hat{F} \tag{4.8}$$

then

$$\bar{A} + \bar{B}\bar{F} = \begin{pmatrix} \mathbf{E}_1 \\ \mathbf{a}_1^T + \mathbf{f}_1^T \\ \mathbf{E}_2 \\ \vdots \\ \mathbf{a}_m^T + \mathbf{f}_m^T \end{pmatrix}$$

where

$$\hat{F} = \begin{pmatrix} \mathbf{f}_1^T \\ \mathbf{f}_2^T \\ \vdots \\ \mathbf{f}_m^T \end{pmatrix}$$

If $\hat{F}$ is determined so that

$$\det(sI - A - BF) = s^n + \gamma_{n-1}s^{n-1} + \ldots + \gamma_0$$

then the $\mathbf{f}$'s are given for $i = 1, 2, 3, 4, \ldots, m-1$ as follows:

$$\left.\begin{aligned} \mathbf{f}_i^{\mathrm{T}} &= -\mathbf{a}_i^{\mathrm{T}} + [\underbrace{0, \ldots, 0}_{\sum_{j=1}^{i} \sigma_j}, 1, 0, \ldots, 0] \\ \mathbf{f}_m^{\mathrm{T}} &= -\mathbf{a}_m^{\mathrm{T}} + [-\gamma_0, -\gamma_1, \ldots, -\gamma_{m-1}] \end{aligned}\right\} \tag{4.9}$$

This $\bar{F}$ gives

$$\bar{A} + \bar{B}\bar{F} = \begin{pmatrix} 0 & 1 & & 0 \\ \vdots & & \ddots & \\ 0 & & 0 & 1 \\ -\gamma_0 & & \ldots, & -\gamma_{m-1} \end{pmatrix}$$

whose characteristic equation is specified a priori. This indicates that the feedback matrix

$$F = \hat{B}_m^{-1} \hat{F} T^{-1} \tag{4.9$'$}$$

makes the characteristic equation of $A + BF$ of the desirable form.

Example 4.3 Find the control law $\mathbf{u} = F\mathbf{x}$ for the system

$$\dot{\mathbf{x}} = \begin{pmatrix} 0 & 0 & 0 & 1 \\ 1 & 0 & 0 & -2 \\ -22 & -11 & -4 & 0 \\ -23 & -6 & 0 & -6 \end{pmatrix} \mathbf{x} + \begin{pmatrix} 0 & 0 \\ 0 & 0 \\ 0 & 1 \\ 1 & 3 \end{pmatrix} \mathbf{u}$$

$$\mathbf{y} = \begin{pmatrix} 0 & 0 & 0 & 1 \\ 0 & 0 & 1 & 0 \end{pmatrix} \mathbf{x}$$

so that poles of the closed loop system are -1, $-1 \pm j$, -2. From Example 3.1, the controllable canonical form of the system is given by

$$\dot{\bar{\mathbf{x}}} = \begin{pmatrix} 0 & 1 & 0 & 0 \\ 0 & 0 & 1 & 0 \\ -6 & -11 & -6 & 0 \\ -11 & 0 & 0 & -4 \end{pmatrix} \bar{\mathbf{x}} + \begin{pmatrix} 0 & 0 \\ 0 & 0 \\ 1 & 3 \\ 0 & 1 \end{pmatrix} \mathbf{u}$$

where $\mathbf{x}$ is related to $\bar{\mathbf{x}}$ by

$$\mathbf{x} = \begin{pmatrix} 0 & 1 & 0 & 0 \\ 1 & -2 & 0 & 0 \\ 0 & 0 & 0 & 1 \\ 0 & 0 & 1 & 0 \end{pmatrix} \bar{\mathbf{x}}$$

Since the desirable characteristic equation is

$$\begin{aligned} \det(sI - \bar{A} - \bar{B}\bar{F}) &= (s+1)(s+1-j)(s+1+j)(s+2) \\ &= (s^2 + 3s + 2)(s^2 + 2s + 2) \\ &= s^4 + 5s^2 + 10s^2 + 10s + 4 \end{aligned}$$

and $\sigma_1 = 3, \sigma_2 = 1$, $\bar{F}$ is given from (4.8) by

$$\bar{F} = \begin{pmatrix} 1 & 3 \\ 0 & 1 \end{pmatrix}^{-1} \begin{pmatrix} 6 & 11 & 6 & 1 \\ 11-4 & 10 & -10 & 4-5 \end{pmatrix}$$

$$= \begin{pmatrix} -15 & 41 & 36 & 4 \\ 7 & -10 & -10 & -1 \end{pmatrix}$$

therefore

$$F = \begin{pmatrix} -15 & 41 & 36 & 4 \\ 7 & -10 & -10 & -1 \end{pmatrix} \begin{pmatrix} 2 & 1 & 0 & 0 \\ 1 & 0 & 0 & 0 \\ 0 & 0 & 0 & 1 \\ 0 & 0 & 1 & 0 \end{pmatrix}$$

$$= \begin{pmatrix} 11 & -15 & 4 & 36 \\ 4 & 7 & -1 & -10 \end{pmatrix}$$

Theorem 4.1 shows that a control law can be determined for a controllable system to provide it with any desirable characteristic equation. Similar results exist for the observable system. If (A, C) is observable, it is known that $(A^{\mathrm{T}}, C^{\mathrm{T}})$ is controllable, therefore there exists a K so that the characteristic equation of $A + KC$

$$\det(sI - A^{\mathrm{T}} - C^{\mathrm{T}}K^{\mathrm{T}}) = \det(sI - A - KC) = 0$$

has any arbitrary symmetric poles.

Corollary

If (A, C) is observable there exists a K for predetermined α_i satisfying

$$\det(Is - A - KC) = s^n + \alpha_{n-1}s^{n-1} + \cdots + \alpha_0$$

4.1.3 Pole Assignment in the Controllable Subspace

In the previous section it was shown that if a system is completely controllable a feedback matrix F can be determined so that the characteristic equation is of a desirable form. In this section we discuss pole assignment for a system which is not controllable.

By Theorem 4.1, it is shown that the controllable subspace of (A, B, C) represented by $\mathcal{R}(\mathcal{C}_1)$ is equal to that of the feedback system $(A + BF, B, C)$ represented by $\mathcal{R}(\mathcal{C}_2)$, $\mathcal{R}(\mathcal{C}_1)$ is A-invariant as shown in Section 2.2.1. Similarly $\mathcal{R}(\mathcal{C}_2)$ is $(A + BF)$-invariant. Therefore $\mathcal{R}(\mathcal{C}_1)$ is $(A + BF)$-invariant and

$$(A + BF)\mathcal{R}(\mathcal{C}_1) \subset \mathcal{R}(\mathcal{C}_1) \tag{4.10}$$

Let the rank of $\mathcal{C}_1$ be n_c and let the basis of $\mathcal{R}(\mathcal{C}_1)$ $(= \mathcal{R}(\mathcal{C}_2))$ be $\mathbf{v}_1, \ldots, \mathbf{v}_{n_c}$. Then from Theorem 2.6, there exists a controllable $(\bar{A}_c, \bar{B}_c)$ for A, B satisfying

$$(A + BF)[\mathbf{v}_1, \ldots, \mathbf{v}_{n_c}] = [\mathbf{v}_1, \ldots, \mathbf{v}_{n_c}](\bar{A}_c + \bar{B}_c\bar{F}_c) \qquad (4.10)'$$

where

$$\bar{F}_c = F[\mathbf{v}_1, \mathbf{v}_2, \dots, \mathbf{v}_{n_c}]$$

Since $(\bar{A}_c, \bar{B}_c)$ is controllable, $\bar{F}_c$ can be determined to satisfy

$$\det(Is - \bar{A}_c - \bar{B}_c\bar{F}_c) = s^{n_c} + \gamma_{n_c-1}s^{n_c-1} + \cdots + \gamma_0$$

for the given γ_i. Let

$$\phi_c(s) = s^{n_c} + \gamma_{n_c-1}s^{n_c-1} + \cdots + \gamma_0$$

then the Cayley-Hamilton theorem yields

$$\phi_c(\bar{A}_c + \bar{B}_c\bar{F}_c) = 0 \tag{4.11}$$

On the other hand, (4.10)′ gives

$$(A + BF)^i[\mathbf{v}_1, \dots, \mathbf{v}_{n_c}] = [\mathbf{v}_1, \dots, \mathbf{v}_{n_c}](\bar{A}_c + \bar{B}_c\bar{F}_c)^i \qquad (i = 0, 1, 2, 3, \dots) \tag{4.12}$$

and therefore (4.12) and (4.11) yield

$$\phi_c(A + BF)[\mathbf{v}_1, \dots, \mathbf{v}_{n_c}] = [\mathbf{v}_1, \dots, \mathbf{v}_{n_c}]\phi_c(\bar{A}_c + \bar{B}_c\bar{F}_c) = 0 \tag{4.13}$$

In section 2.2.3, it was shown that the general solution of

$$\dot{\mathbf{x}} = (A + BF)\mathbf{x}$$

is given by (2.49) for an initial condition in a certain subspace. And for $\mathbf{x}(0) \in \mathscr{C}$ satisfying $\phi_c(A + BF)\mathscr{C} = 0$, $\mathbf{x}(t)$ is given by a linear combination of (2.49) corresponding to the roots of $\phi_c(s) = 0$.

Theorem 4.3

There exists a feedback matrix F to realize the given characteristic equation $\phi_c(s)$ corresponding to the controllable subspace $\mathscr{X}_c$ of (A, B) satisfying $\phi_c(A + BF)\mathscr{X}_c = 0$.

When some characteristic roots of A have positive real parts, $\phi^+(s)$ denotes the polynomial which has these characteristic roots. Then for any initial condition

$$\mathbf{x}(0) \in \mathscr{X}^+(A)$$

where

$$\mathscr{X}^+(A) = \mathscr{N}[\phi^+(A)],$$

then the solution of

$$\dot{\mathbf{x}} = A\mathbf{x}$$

is known to diverge as we have seen in Chapter 2. Therefore if

$$\mathscr{X}^+(A) \subset \mathscr{R}(\mathscr{C}) \tag{4.14}$$

there exists a feedback matrix which makes the closed loop system stable. So if (4.14) is satisfied, the system is said to be 'stabilizable' or alternatively (A, B) is said to be 'stabilizable'. When $(A^{\mathrm{T}}, C^{\mathrm{T}})$ is stabilizable, (A, C) is said to be 'detectable'. Thus 'detectability' is the dual concept of 'stabilizability'.

Example 4.4 Check whether the following system is stabilizable or not

$$\dot{\mathbf{x}} = \begin{pmatrix} 0 & -2 & 2 \\ 1 & 3 & 1 \\ 0 & 0 & -1 \end{pmatrix} \mathbf{x} + \begin{pmatrix} 0 \\ 1 \\ 0 \end{pmatrix} u$$

The controllable subspace of the system is

$$\mathscr{R}\begin{pmatrix} 0 & -2 & -6 \\ 1 & 3 & 7 \\ 0 & 0 & 0 \end{pmatrix} = \left\{ \begin{pmatrix} 0 \\ 1 \\ 0 \end{pmatrix}, \begin{pmatrix} -2 \\ 3 \\ 0 \end{pmatrix} \right\}$$

Since the characteristic roots of A are $-1, 1, 2$,

$$\begin{aligned} \mathscr{X}^{+}(A) &= \mathscr{N}(I - A)(2I - A) \\ &= \mathscr{N}(2I - 3A + A^2) \\ &= \mathscr{N}\begin{pmatrix} 0 & 0 & -10 \\ 0 & 0 & 1 \\ 0 & 0 & 6 \end{pmatrix} = \left\{ \begin{pmatrix} 1 \\ 0 \\ 0 \end{pmatrix}, \begin{pmatrix} 0 \\ 1 \\ 0 \end{pmatrix} \right\}, \\ &\subset \mathscr{R}(\mathscr{C}) \end{aligned}$$

So the system is found to be stabilizable.

4.2 THE DECOUPLING PROBLEM

4.2.1 Decoupling by State Feedback

A system with m inputs and m outputs is said to be decoupled if and only if its transfer function is given by

$$H(s) = \begin{pmatrix} h_{11}(s) & & & \\ & h_{22}(s) & & 0 \\ & & \ddots & \\ 0 & & & h_{mm}(s) \end{pmatrix} = \mathrm{diag}(h_{11}(s), h_{22}(s), \ldots, h_{mm}(s)) \tag{4.15}$$

where $h_{ii}(s)$ is not zero. For a decoupled system the effect of the ith input appears only in the ith output. In this section we consider decoupling of the system

$$\dot{\mathbf{x}} = A\mathbf{x} + B\mathbf{u} \qquad \mathbf{x}(0) = \mathbf{x}_0 \tag{4.16a}$$

$$\mathbf{y} = C\mathbf{x} \tag{4.16b}$$

where $\mathbf{u}, \mathbf{y} \in R^m, \mathbf{x} \in R^n, A \in R^{n \times n}, B \in R^{n \times m}, C \in R^{m \times n}$ by the control law

$$\mathbf{u} = F\mathbf{x} + G\mathbf{v}. \tag{4.17}$$

This problem is called decoupling by state feedback.

When the control law (4.17) is employed, the resultant system becomes

$$\dot{\mathbf{x}} = (A + BF)\mathbf{x} + BG\mathbf{v} \tag{4.18a}$$

$$\mathbf{y} = C\mathbf{x} \tag{4.18b}$$

and the transfer function matrix is given by

$$H_{FG}(s) = C(sI - A - BF)^{-1}BG \tag{4.18c}$$

Therefore decoupling by state feedback requires one to find the control matrices F and G which make $H_{FG}(s)$ diagonal and non-singular.

Before treating the decoupling problem we first consider the relation between the transfer function $H_{FG}(s)$ and that of (4.16). The transfer function matrix of (4.16) is given by

$$H(s) = C(sI - A)^{-1}B \tag{4.16c}$$

Proposition 4.1

The transfer function matrix $H_{FG}(s)$ of (4.18c) is related to $H(s)$ of (4.16c) by

$$\begin{aligned} H_{FG}(s) &= H(s)[I + F(sI - A - BF)^{-1}B]G \\ &= H(s)[I - F(sI - A)^{-1}B]^{-1}G \end{aligned} \tag{4.19}$$

Proof

$$\begin{aligned} H_{FG}(s) &= C(sI - A - BF)^{-1}BG \\ &= C(sI - A)^{-1}[(sI - A - BF) + BF](sI - A - BF)^{-1}BG \\ &= H(s)B^{-1}[I + BF(sI - A - BF)^{-1}]BG \\ &= H(s)[I + F(sI - A - BF)^{-1}B]G \end{aligned}$$

Now consider

$$\begin{aligned} &[I + F(sI - A - BF)^{-1}B][I - F(sI - A)^{-1}B] \\ &\quad = I + F(sI - A - BF)^{-1}(sI - A - BF + BF)(sI - A)^{-1}B \\ &\qquad - F(sI - A)^{-1}B - F(sI - A - BF)^{-1}BF(sI - A)^{-1}B \\ &\quad = I \end{aligned}$$

Therefore

$$[I + F(sI - A - BF)^{-1}B] = [I - F(sI - A)^{-1}B]^{-1} \tag{4.20}$$

and (4.19) is proved.

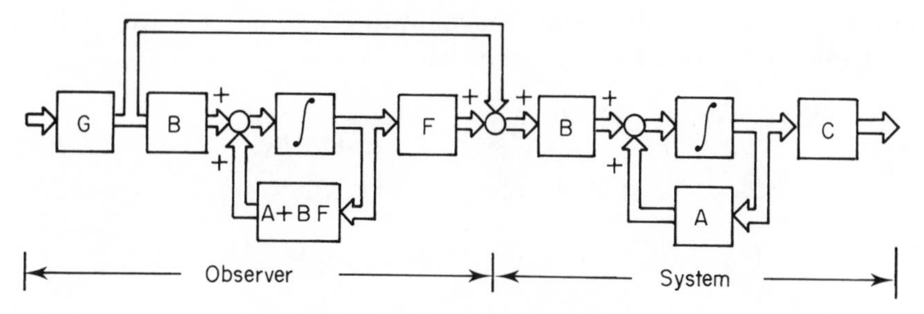

Figure 4.2 Series compensator

Proposition 4.1 indicates that controlling the system (4.16) with the control law (4.17) is equivalent to compensating the system (4.16) serially using the compensator

$$H_c(s) = [I + F(sI - A - BF)^{-1}B]G \tag{4.21}$$

This compensator can be represented by the state space equations

$$\dot{\mathbf{x}}_c = (A + BF)\mathbf{x}_c + BG\mathbf{v} \tag{4.22a}$$

$$\mathbf{y}_c = F\mathbf{x}_c + G\mathbf{v} \tag{4.22b}$$

where $\mathbf{v}$ is the input to the compensator and $\mathbf{y}_c$ is its output. The original feedback control system is shown in Figure 4.1 and its equivalent representation with the series precompensator is shown in Figure 4.2. Using this proposition the necessary and sufficient condition for the system to be decoupled by state feedback control is given by the following theorem.

Theorem 4.4

There exists a control law of the form (4.17) to decouple the system (4.16) if and only if the matrix

$$B^* = \begin{pmatrix} \mathbf{c}_1^{\mathrm{T}} A^{\sigma_1 - 1} B \\ \mathbf{c}_2^{\mathrm{T}} A^{\sigma_2 - 1} B \\ \vdots \\ c_m^{\mathrm{T}} A^{\sigma_m - 1} B \end{pmatrix} \tag{4.23}$$

is non-singular, where

$$C^{\mathrm{T}} = [\mathbf{c}_1, \mathbf{c}_2, \ldots, \mathbf{c}_m]$$

and σ_i $(i = 1, 2, \ldots, m)$ is an integer given by

$$\sigma_i = \begin{cases} \min(j \mid \mathbf{c}_i^{\mathrm{T}} A^{j-1} B \neq \mathbf{0}^{\mathrm{T}}) \\ n - 1; \mathbf{c}_i^{\mathrm{T}} A^j B = \mathbf{0}^{\mathrm{T}}, \forall_j \end{cases} \tag{4.24}$$

Proof Necessity is proved first. The ith row of the transfer function $H(s)$ of (4.16) can be expanded in polynomials of s^{-i} as

$$\begin{aligned}\mathbf{h}(s)_i^{\mathrm{T}} &= \mathbf{c}_i^{\mathrm{T}}(sI-A)^{-1}B = \mathbf{c}_i^{\mathrm{T}}Bs^{-1} + \mathbf{c}_i^{\mathrm{T}}ABs^{-2} + \cdots \\ &= s^{-\sigma_i}(\mathbf{c}_i^{\mathrm{T}}A^{\sigma_i-1}B + \mathbf{c}_i^{\mathrm{T}}A^{\sigma_i}Bs^{-1} + \mathbf{c}_i^{\mathrm{T}}A^{\sigma_i+1}Bs^{-2} + \cdots) \\ &= s^{-\sigma_i}\{\mathbf{c}_i^{\mathrm{T}}A^{\sigma_i-1}B + \mathbf{c}_i^{\mathrm{T}}A^{\sigma_i}(sI-A)^{-1}B\} \end{aligned} \tag{4.25}$$

Using B^* of (4.23) $H(s)$ can be written as

$$H(s) = \begin{pmatrix} s^{-\sigma_1} & & & \\ & s^{-\sigma_2} & & 0 \\ & 0 & \ddots & \\ & & & s^{-\sigma_m} \end{pmatrix} [B^* + C^*(sI-A)^{-1}B] \tag{4.26}$$

where C^* is

$$C^* = \begin{pmatrix} \mathbf{c}_1^{\mathrm{T}}A^{\sigma_1} \\ \mathbf{c}_2^{\mathrm{T}}A^{\sigma_2} \\ \vdots \\ \mathbf{c}_m^{\mathrm{T}}A^{\sigma_m} \end{pmatrix} \tag{4.27}$$

Thus from Proposition 4.1 the transfer function matrix of the feedback system $H_{FG}(s)$ is given by

$$\begin{aligned} H_{FG}(s) &= H(s)[I + F(sI-A-BF)^{-1}B]G \\ &= \mathrm{diag}(s^{-\sigma_1}, s^{-\sigma_2}, \ldots, s^{-\sigma_m})[B^* + C^*(sI-A)^{-1}B] \\ &\qquad \times [I + F(sI-A-BF)^{-1}B]G \end{aligned} \tag{4.28}$$

If $H_{FG}(s)$ is to be non-singular and diagonal then $[B^* + C^*(sI-A)^{-1}B]$ $[I + F(sI-A-BF)^{-1}B]G$ should be non-singular and diagonal. Thus G should be non-singular and the coefficient matrix for s^0 of the polynomial matrix B^*G should be diagonal. Since G is non-singular and all row vectors of B^* are non-zero, B^*G is non-singular. Thus the non-singularity of B^* is a necessary condition for decoupling.

The sufficiency is proved by constructing the control law to decouple the system as follows. If B^* is non-singular, let F and G be chosen as

$$G = B^{*-1} \tag{4.29a}$$

$$F = -B^{*-1}C^* \tag{4.29b}$$

Now from Proposition 4.1,

$$\begin{aligned} H_{FG}(s) = \mathrm{diag}(s^{-\sigma_1}, s^{-\sigma_2}, \ldots, s^{-\sigma_m})&[B^* + C^*(sI-A)^{-1}B] \\ &\times [G^{-1} - G^{-1}F(sI-A)^{-1}B]^{-1}. \end{aligned} \tag{4.30}$$

Substituting (4.29) into (4.30) gives

$$\begin{aligned} H_{FG}(s) &= \text{diag}(s^{-\sigma_1}, s^{-\sigma_2}, \ldots, s^{-\sigma_m})[B^* + C^*(sI - A)^{-1}B] \\ &\quad \times [B^* + C^*(sI - A)^{-1}B]^{-1} \\ &= \text{diag}(s^{-\sigma_1}, s^{-\sigma_2}, \ldots, s^{-\sigma_m}) \end{aligned} \tag{4.31}$$

and the system is decoupled.

This form of decoupled system having $s^{-\sigma_i}$ in the diagonal element as (4.31) is called an integrator decoupled system.

Example 4.5 The system

$$\dot{\mathbf{x}} = \begin{pmatrix} 0 & 0 & 0 \\ 0 & 0 & 1 \\ -1 & -2 & -3 \end{pmatrix} \mathbf{x} + \begin{pmatrix} 1 & 0 \\ 0 & 0 \\ 0 & 1 \end{pmatrix} \mathbf{u}$$

$$\mathbf{y} = \begin{pmatrix} 1 & 1 & 0 \\ 0 & 0 & 1 \end{pmatrix} \mathbf{x}$$

is required to be integrator decoupled by the control law (4.17).

Solution The transfer function matrix of the system is calculated as

$$H(s) = \begin{pmatrix} 1 & 1 & 0 \\ 0 & 0 & 1 \end{pmatrix} \begin{pmatrix} s & 0 & 0 \\ 0 & s & -1 \\ 1 & 2 & s+3 \end{pmatrix}^{-1} \begin{pmatrix} 1 & 0 \\ 0 & 0 \\ 0 & 1 \end{pmatrix}$$

$$= \begin{pmatrix} \dfrac{s^2 + 3s + 1}{s(s+1)(s+2)} & \dfrac{1}{(s+1)(s+2)} \\ -\dfrac{1}{(s+1)(s+2)} & \dfrac{s}{(s+1)(s+2)} \end{pmatrix}$$

and its internal structure is depicted in Figure 4.3.

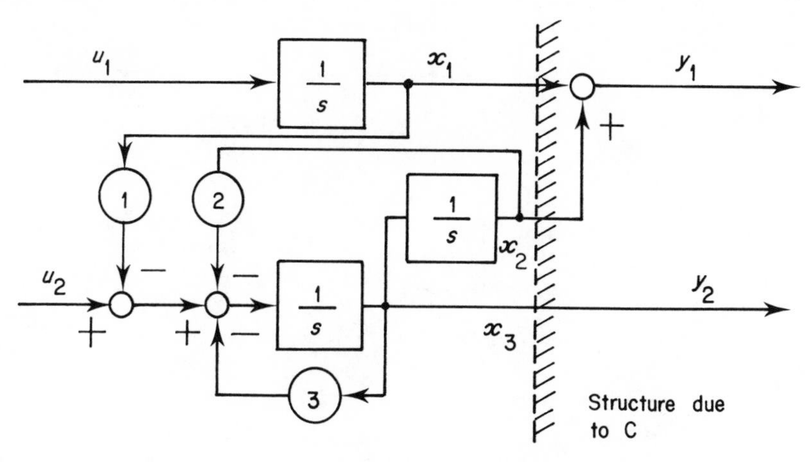

Figure 4.3 System of Example 4.5

Now

$$\mathbf{c}_1^{\mathrm{T}} B = (1, 0) \text{ which yields } \sigma_1 = 1 \text{ and } \mathbf{c}_1^{\mathrm{T}} A = (0, 0, 1)$$

and

$$\mathbf{c}_2^{\mathrm{T}} B = (0, 1) \text{ which yields } \sigma_2 = 1 \text{ and } \mathbf{c}_2^{\mathrm{T}} A = (-1, -2, -3),$$

therefore B^* of (4.23) and C^* of (4.27) are

$$B^* = \begin{pmatrix} 1 & 0 \\ 0 & 1 \end{pmatrix}, \quad C^* = \begin{pmatrix} 0 & 0 & 1 \\ -1 & -2 & -3 \end{pmatrix}$$

Since B^* is non-singular, the system can be decoupled. From (4.29) the control law has

$$G = B^{*-1} = I, \text{ and } F = -B^{*-1}C^* = \begin{pmatrix} 0 & 0 & -1 \\ 1 & 2 & 3 \end{pmatrix}.$$

Thus the system becomes

$$\dot{\mathbf{x}} = \begin{pmatrix} 0 & 0 & -1 \\ 0 & 0 & 1 \\ 0 & 0 & 0 \end{pmatrix} \mathbf{x} + \begin{pmatrix} 1 & 0 \\ 0 & 0 \\ 0 & 1 \end{pmatrix} \mathbf{v}$$

$$\mathbf{y} = \begin{pmatrix} 1 & 1 & 0 \\ 0 & 0 & 1 \end{pmatrix} \mathbf{x}$$

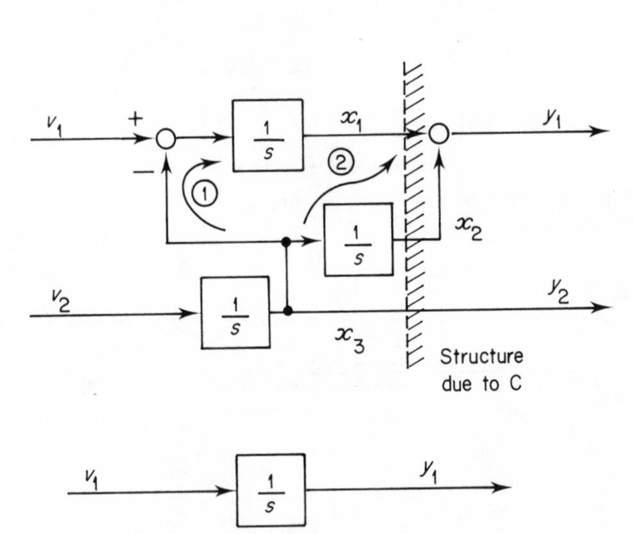

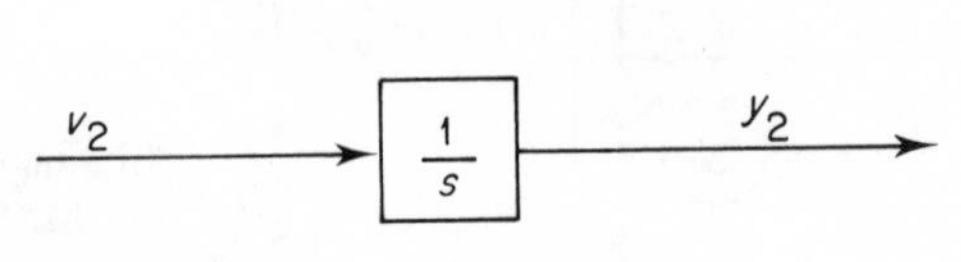

Figure 4.4

which has the transfer function

$$H_{\mathrm{FG}}(s) = \begin{pmatrix} 1 & 1 & 0 \\ 0 & 0 & 1 \end{pmatrix} \begin{pmatrix} s & 0 & 1 \\ 0 & s & -1 \\ 0 & 0 & s \end{pmatrix}^{-1} \begin{pmatrix} 1 & 0 \\ 0 & 0 \\ 0 & 1 \end{pmatrix}$$

$$= \begin{pmatrix} \frac{1}{s} & 0 \\ 0 & \frac{1}{s} \end{pmatrix}$$

The system is therefore integrator decoupled and its structure is depicted in Figure 4.4. As can be seen from (4.18) the control law does not change the output matrix C. The measurement structure is therefore unchanged, and decoupling is achieved by cancelling the effect of path ② by path ①.

4.2.2 Pole Assignment and Zeros of Decoupled Systems

In this section, the control law matrices F and G are determined so that the decoupled system has preassigned poles. This is given by the following theorem.

Theorem 4.5

When the system (4.16) can be decoupled by state feedback and F and G are determined from

$$G = B^{*-1}$$
$$(-G^{-1}F)_i = \mathbf{c}_i^{\mathrm{T}} A^{\sigma_i} + \alpha_{i1}\mathbf{c}_i^{\mathrm{T}} A^{\sigma_i - 1} + \alpha_{i2}\mathbf{c}_i^{\mathrm{T}} A^{\sigma_i - 2} + \cdots + \alpha_{i\sigma_i}\mathbf{c}_i^{\mathrm{T}} \tag{4.32}$$

where $(-G^{-1}F)_i$ denotes the ith row vector of $-G^{-1}F$, then the resultant feedback system has the transfer function given by

$$H_{FG}(s) = \begin{pmatrix} \dfrac{1}{s^{\sigma_1} + \alpha_{11}s^{\sigma_1 - 1} + \cdots + \alpha_{1\sigma_1}}, & & 0 \\ & \ddots & \\ 0 & & \dfrac{1}{s^{\sigma_m} + \alpha_{m1}s^{\sigma_m - 1} + \cdots + \alpha_{m\sigma_m}} \end{pmatrix} \tag{4.33}$$

Proof Let the ith row vector of the transfer function matrix of (4.16) be denoted by $\mathbf{h}(s)_i^{\mathrm{T}}$ and multiply by $(s^{\sigma_i} + \cdots + \alpha_{i\sigma_i})$, then from (4.25)

$$
\begin{aligned}
(s^{\sigma_i} + \alpha_{i1}s^{\sigma_i - 1} + \cdots + \alpha_{i\sigma_i})\mathbf{h}(s)_i^{\mathrm{T}} \\
&= \mathbf{c}_i^{\mathrm{T}} A^{\sigma_i - 1}B + (\mathbf{c}_i^{\mathrm{T}} A^{\sigma_i} + \alpha_{i1}\mathbf{c}_i^{\mathrm{T}} A^{\sigma_i - 1} + \cdots + \alpha_{i\sigma_i}\mathbf{c}_i^{\mathrm{T}})Bs^{-1} \\
&\quad + \cdots + (\mathbf{c}_i^{\mathrm{T}} A^{\sigma_i} + \alpha_{i1}\mathbf{c}_i^{\mathrm{T}} A^{\sigma_i - 1} + \cdots + \alpha_{1\sigma_i}\mathbf{c}_i^{\mathrm{T}})A^{j-1}Bs^{-j} \\
&\quad + \cdots \\
&= \mathbf{c}_i^{\mathrm{T}} A^{\sigma_i - 1}B + (\mathbf{c}_i^{\mathrm{T}} A^{\sigma_i} + \alpha_{i1}\mathbf{c}_i^{\mathrm{T}} A^{\sigma_i - 1} + \cdots + \alpha_{i\sigma_i}\mathbf{c}_i^{\mathrm{T}})(sI - A)^{-1}B \\
&= \mathbf{c}_i^{\mathrm{T}} A^{\sigma_i - 1}B + \mathbf{c}_i^{**\mathrm{T}}(sI - A)^{-1}B \qquad (4.34)
\end{aligned}
$$

where

$$
\mathbf{c}_i^{**\mathrm{T}} = (\mathbf{c}_i^{\mathrm{T}} A^{\sigma_i} + \alpha_{i1}\mathbf{c}_i^{\mathrm{T}} A^{\sigma_i - 1} + \cdots + \alpha_{i\sigma_i}\mathbf{c}_i^{\mathrm{T}}) \qquad (4.35)
$$

Dividing both sides of (4.34) by $(s^{\sigma_i} + \alpha_{i1}s^{\sigma_i - 1} + \cdots + \alpha_{i\sigma_i})$, $H(s)$ is given by

$$
\begin{aligned}
H(s) = \mathrm{diag}[&(s^{\sigma_1} + \alpha_{11}s^{\sigma_1 - 1} + \cdots + \alpha_{1\sigma_1})^{-1}, (s^{\sigma_2} + \alpha_{21}s^{\sigma_2 - 1} + \cdots + \alpha_{2\sigma_2})^{-1}, \\
&\ldots, (s^{\sigma_m} + \alpha_{m1}s^{\sigma_m - 1} + \cdots + \alpha_{m\sigma_m})^{-1}][B^* + C^{**}(sI - A)^{-1}B] \qquad (4.36)
\end{aligned}
$$

where

$$
C^{**} = \begin{pmatrix} \mathbf{c}_1^{**\mathrm{T}} \\ \mathbf{c}_2^{**\mathrm{T}} \\ \vdots \\ c_m^{**\mathrm{T}} \end{pmatrix}
$$

Then from (4.19) and (4.36)

$$
\begin{aligned}
H_{FG}(s) = \mathrm{diag}[&(s^{\sigma_1} + \alpha_{11}s^{\sigma_1 - 1} + \cdots + \alpha_{1\sigma_1})^{-1}, \ldots, (s^{\sigma_m} + \alpha_{m1}s^{\sigma_m - 1} + \cdots \\
&+ \alpha_{m\sigma_m})^{-1}][B^* + C^{**}(sI - A)^{-1}B][G^{-1} - G^{-1}F(sI - A)^{-1}B]^{-1} \qquad (4.37)
\end{aligned}
$$

Thus by choosing G and F as (4.32), the transfer function matrix of the closed loop system is given by (4.33).

Example 4.6 In Example 4.5, the integrator decoupled system is obtained. In this example, the same system is required to be decoupled to give

$$
H_{FG}(s) = \mathrm{diag}[(s + 1)^{-1}, (s + 2)^{-1}]
$$

From Example 4.5, $B^* = I$ and therefore $G = I$. Since $\alpha_{11} = 1, \alpha_{21} = 2$, (4.32) gives

$$
\begin{aligned}
-F &= \begin{pmatrix} \mathbf{c}_1^{\mathrm{T}} A \\ \mathbf{c}_2^{\mathrm{T}} A \end{pmatrix} + \begin{pmatrix} \alpha_{11}\mathbf{c}_1^{\mathrm{T}} \\ \alpha_{21}\mathbf{c}_2^{\mathrm{T}} \end{pmatrix} \\
&= \begin{pmatrix} 0 & 0 & 1 \\ -1 & -2 & -3 \end{pmatrix} + \begin{pmatrix} 1 & 1 & 0 \\ 0 & 0 & 2 \end{pmatrix} = \begin{pmatrix} 1 & 1 & 1 \\ -1 & -2 & -1 \end{pmatrix}
\end{aligned}
$$

$$A + BF = \begin{pmatrix} 0 & 0 & 0 \\ 0 & 0 & 1 \\ -1 & -2 & -3 \end{pmatrix} + \begin{pmatrix} 1 & 0 \\ 0 & 0 \\ 0 & 1 \end{pmatrix} \begin{pmatrix} -1 & -1 & -1 \\ 1 & 2 & 1 \end{pmatrix}$$

$$= \begin{pmatrix} -1 & -1 & -1 \\ 0 & 0 & 1 \\ 0 & 0 & -2 \end{pmatrix}.$$

This yields $H_{FG}(s) = \mathrm{diag}[(s+1)^{-1}, (s+2)^{-1}]$.

In the above example, a 3rd order system has been decoupled to give a 2nd order system without finite zeros. Thus a pole-zero cancellation has taken place. In the next theorem the zero positions are considered in the decoupling.

Theorem 4.6

The linear system (4.1) can be decoupled by the feedback control law (4.17) with the ith diagonal element of the transfer function of the decoupled system having a numerator $n_i(s)$ if and only if the ith row of $H(s)$, denoted by $\mathbf{h}_i^{\mathrm{T}}(s)$ is represented by

$$\mathbf{h}(s)_i^{\mathrm{T}} = \mathbf{c}_i^{\mathrm{T}}(sI - A)^{-1}B = n_i(s)\bar{\mathbf{c}}_i^{\mathrm{T}}(sI - A)^{-1}B \qquad (i = 1, \ldots, m) \qquad (4.38)$$

and

$$\bar{B}^* = \begin{pmatrix} \bar{\mathbf{c}}_1^{\mathrm{T}} A^{\bar{\sigma}_1 - 1} B \\ \bar{\mathbf{c}}_2^{\mathrm{T}} A^{\bar{\sigma}_2 - 1} B \\ \vdots \\ \bar{\mathbf{c}}_m^{\mathrm{T}} A^{\bar{\sigma}_m - 1} B \end{pmatrix} \qquad (4.23)'$$

is non-singular, where $\bar{\sigma}_i$ is

$$\bar{\sigma}_i = \begin{cases} \min(j \mid \bar{\mathbf{c}}_i^{\mathrm{T}} A^{j-1} B \neq \mathbf{0}^{\mathrm{T}}) \\ n - 1: \bar{\mathbf{c}}_i^{\mathrm{T}} A^{j-1} B = \mathbf{0}^{\mathrm{T}}, \forall j \end{cases}$$

Proof Sufficiency is proved first. Let

$$\bar{C} = \begin{pmatrix} \bar{\mathbf{c}}_1^{\mathrm{T}} \\ \vdots \\ \bar{\mathbf{c}}_m^{\mathrm{T}} \end{pmatrix}$$

then from (4.38), the transfer function matrix $H(s)$ is

$$H(s) = \mathrm{diag}[n_1(s), n_2(s), \ldots, n_m(s)]\bar{C}(sI - A)^{-1}B \qquad (4.39)$$

From Theorem 4.5 and (4.23)′, $(A, B, \bar{C})$ can be decoupled to give

$$H_{FG}(s) = \mathrm{diag}[n_1(s), n_2(s), \ldots, n_m(s)]\mathrm{diag}[d_1^{-1}(s), \ldots, d_m^{-1}(s)]$$

Therefore the ith component of the decoupled transfer function $h_{ii}(s)$ is represented by

$$h_{ii}(s) = n_i(s)/d_i(s) \qquad i = 1, 2, \ldots, m \tag{4.40}$$

where $d_i(s)$ is determined not to have common factors with $n_i(s)$. Thus sufficiency is proved.

Necessity is proved next. When the diagonal element of the decoupled system is given by (4.40), the ith row vector of the decoupled transfer function $\mathbf{h}_{FG}(s)_i^{\mathrm{T}}$ is

$$\begin{aligned}\mathbf{h}_{FG}(s)_i^{\mathrm{T}} &= \mathbf{c}_i^{\mathrm{T}}(sI - A - BF)^{-1}BG \\ &= n_i(s)/d_i(s)\mathbf{e}_i^{\mathrm{T}}\end{aligned} \tag{4.41}$$

From (3.23) and (4.41), there exists a $\bar{\mathbf{c}}_i^{\mathrm{T}}$ satisfying

$$\mathbf{h}_{FG}(s)_i^{\mathrm{T}} = n_i(s)\bar{\mathbf{c}}_i^{\mathrm{T}}(sI - A - BF)^{-1}BG \tag{4.41$'$}$$

On the other hand from (4.19), (4.41) and (4.41)$'$ it is easily shown that

$$\begin{aligned}&\mathbf{c}_i^{\mathrm{T}}(sI - A)^{-1}B[I - F(sI - A)^{-1}B]^{-1}G \\ &\qquad = n_i(s)\bar{\mathbf{c}}_i^{\mathrm{T}}(sI - A)^{-1}B[I - F(sI - A)^{-1}B]^{-1}G\end{aligned}$$

(4.39) is then derived by dividing both sides of the above by $[I - F(sI - A)^{-1}B]^{-1}G$ from the right side.

From the above theorem, it is found that the numerator terms in the decoupled system are given by the properties of the original system and cannot be designed arbitrarily. In the next section an algorithm to determine the numerator terms using Wolovich and Falb's structure theorem is presented.

The Algorithm to Obtain the Numerator Terms of Decoupled Systems

The 1st step: Luenberger's canonical form (A_c, B_c, C_c) is calculated for the given system (A, B, C). Then from (3.23) the transfer function is given by

$$H(s) = C_c S(s)\delta_c^{-1}(s)\hat{B}_m(s) \tag{3.23}$$

where

$$S(s) = \begin{pmatrix} 1 & 0 & \ldots & 0 \\ s & \vdots & & \vdots \\ \vdots & & & \\ s^{\sigma_1 - 1} & & & \\ 0 & & & \\ \vdots & & & s^{\sigma_m - 2} \\ 0 & & & s^{\sigma_m - 1} \end{pmatrix}$$

$$\delta_c(s) = \begin{pmatrix} s^{\sigma_1} & & & \\ & s^{\sigma_2} & 0 & \\ & 0 & \ddots & \\ & & & s^{\sigma_m} \end{pmatrix} - \begin{pmatrix} \mathbf{a}_1^T \\ \mathbf{a}_2^T \\ \vdots \\ \mathbf{a}_m^T \end{pmatrix} S(s)$$

$$\hat{B}_m = \begin{pmatrix} 1 & b_{12} & b_{13} & \dots & b_{1m} \\ 0 & 1 & b_{23} & \dots & b_{2m} \\ \vdots & & & \ddots & \vdots \\ 0 & \dots & & & 1 \end{pmatrix}$$

and $\mathbf{a}_i^T$ and b_{ij} are found from

$$A_c = \begin{pmatrix} A_{11} & A_{12} & \dots & A_{1m} \\ A_{21} & A_{22} & \dots & A_{2m} \\ \vdots & & & \vdots \\ A_{m1} & A_{m2} & \dots & A_{mm} \end{pmatrix}$$

$$[A_{i1}, \dots, A_{ii}, \dots, A_{im}] = \begin{pmatrix} 0, & \begin{matrix} 0 \\ \vdots \\ 0 \end{matrix} & I\sigma_{i-1} & , \dots, 0 \\ \hline & \mathbf{a}_i^T & & \end{pmatrix}$$

$$B_c = \begin{pmatrix} B_1 \\ \vdots \\ B_m \end{pmatrix}, \quad B_i = \begin{pmatrix} 0 & \cdots & & & 0 \\ \vdots & & & & \vdots \\ \vdots & \cdots & & & 0 \\ 0 & 0 & 1 & b_{ii+1} & b_{im} \end{pmatrix}$$

The 2nd step: The $\bar{C}_c$ satisfying

$$C_c S(s)\delta_c^{-1}(s)\hat{B}_m = \operatorname{diag}[n_1(s), \dots, n_m(s)]\bar{C}_c S(s)\delta_c^{-1}(s)\hat{B}_m \qquad (4.42)$$

is determined.

The 3rd step: The control law F_c and G_c to decouple $(A_c, B_c, \bar{C}_c)$ is determined, and if necessary F_c and G_c are transformed to F and G for (A, B, C).

Using the previous example, the given algorithm is illustrated. Since the system is already in the controllable canonical form, the first step is from (3.23),

$$H(s) = \left\{\begin{pmatrix} 1 & 1 & 0 \\ 0 & 0 & 1 \end{pmatrix}\begin{pmatrix} 1 & 0 \\ 0 & 1 \\ 0 & s \end{pmatrix}\right\}\left\{\begin{pmatrix} s & 0 \\ 0 & s^2 \end{pmatrix} + \begin{pmatrix} 0 & 0 & 0 \\ 1 & 2 & 3 \end{pmatrix}\begin{pmatrix} 1 & 0 \\ 0 & 1 \\ 0 & s \end{pmatrix}\right\}^{-1}$$

$$= \begin{pmatrix} 1 & 1 \\ 0 & s \end{pmatrix}\begin{pmatrix} s & 0 \\ 1 & s^2 + 3s + 2 \end{pmatrix}^{-1}$$

The 2nd step:

$$\begin{pmatrix}1 & 1\\0 & s\end{pmatrix}=\begin{pmatrix}1 & 0\\0 & s\end{pmatrix}\begin{pmatrix}1 & 1 & 0\\0 & 1 & 0\end{pmatrix}\begin{pmatrix}1 & 0\\0 & 1\\0 & s\end{pmatrix}$$

yields

$$\bar{C}=\begin{pmatrix}1 & 1 & 0\\0 & 1 & 0\end{pmatrix}$$

so that $n_1(s)=1,\ n_2(s)=s$.
The 3rd step: $\bar{\mathbf{c}}_1^{\mathrm{T}}B=[1\quad 0]$, $\bar{\mathbf{c}}_2^{\mathrm{T}}B=[0\quad 0]$, $\bar{\mathbf{c}}_2^{\mathrm{T}}AB=[0\quad 1]$ yields $B^*=I$. So $G=I$ and from (4.32) F is given by

$$F=\begin{pmatrix}0 & 0 & -1\\1 & 2 & 3\end{pmatrix}+\begin{pmatrix}-\alpha_{11} & -\alpha_{11} & 0\\0 & 0 & -\alpha_{21}\end{pmatrix}+\begin{pmatrix}0 & 0 & 0\\0 & -\alpha_{22} & 0\end{pmatrix}$$

By choosing $\alpha_{11}=1,\ \alpha_{21}=3,\ \alpha_{22}=2,\ F$ is determined as

$$F=\begin{pmatrix}-1 & -1 & -1\\1 & 0 & 0\end{pmatrix}$$

The resultant transfer function is

$$H_{FG}(s)=\begin{pmatrix}1 & 1 & 0\\0 & 0 & 1\end{pmatrix}\begin{pmatrix}s+1 & 1 & 1\\0 & s & -1\\0 & 2 & s+3\end{pmatrix}^{-1}\begin{pmatrix}1 & 0\\0 & 0\\0 & 1\end{pmatrix}$$

$$=\begin{pmatrix}\dfrac{1}{s+1} & 0\\0 & \dfrac{s}{s^2+3s+2}\end{pmatrix}$$

and in this case all the poles are found to be assignable. This, however, is not generally the case for decoupled systems.

PROBLEMS

P4.1 Find feedback $\mathbf{f}^{\mathrm{T}}$ to assign poles $-1,\ -1\pm j$ for the following systems.

(1–1)

$$(A,\mathbf{b})=\left(\begin{bmatrix}0 & 1 & 2\\1 & 3 & 4\\0 & 5 & 6\end{bmatrix},\begin{bmatrix}1\\0\\0\end{bmatrix}\right)$$

(1–2)

$$(A,\mathbf{b})=\left(\begin{bmatrix}3 & -2 & 0\\0 & 0 & 1\\4 & -3 & 0\end{bmatrix},\begin{bmatrix}1\\0\\1\end{bmatrix}\right)$$

(1–3)

$$(A, \mathbf{b}) = \left(\begin{bmatrix} -1 & 0 & 1 \\ 0 & 1 & 0 \\ 0 & -1 & 1 \end{bmatrix}, \begin{bmatrix} 1 \\ -1 \\ 1 \end{bmatrix} \right)$$

P4.2 Find feedback $\mathbf{f}^{\mathrm{T}}$ to assign poles $-1 \pm j$ for the systems

(2–1)

$$(A, \mathbf{b}) = \left(\begin{bmatrix} 1 & 0 \\ -2 & 1 \end{bmatrix}, \begin{bmatrix} -1 \\ 1 \end{bmatrix} \right)$$

(2–2)

$$(A, \mathbf{b}) = \left(\begin{bmatrix} 1 & 0 \\ 0 & -1 \end{bmatrix}, \begin{bmatrix} 1 \\ 1 \end{bmatrix} \right)$$

P4.3 Check whether the systems given in P2.1 are stabilizable or not.

P4.4 Check whether the following systems can be decoupled or not, and if the system can be decoupled find F and G to yield integrator decoupled systems.

(4–1)

$$(A, B, C) = \left(\begin{bmatrix} 0 & 1 & 0 \\ 1 & 1 & 1 \\ 1 & 0 & 0 \end{bmatrix}, \begin{bmatrix} 0 & 1 \\ 1 & 0 \\ 0 & -1 \end{bmatrix}, \begin{bmatrix} 1 & 0 & 1 \\ 0 & 1 & 0 \end{bmatrix} \right)$$

(4–2)

$$(A, B, C) = \left(\begin{bmatrix} 1 & 0 & 1 \\ 0 & 1 & 1 \\ 1 & 1 & 0 \end{bmatrix}, \begin{bmatrix} 1 & 0 \\ 0 & 1 \\ 1 & 1 \end{bmatrix}, \begin{bmatrix} 1 & 1 & 0 \\ 0 & 1 & 1 \end{bmatrix} \right)$$

P4.5 Show that for the n-dimensional system with m inputs and p outputs the output feedback control

$$\mathbf{u} = k\mathbf{f}^{\mathrm{T}}\mathbf{y}$$

can assign p poles by choosing $\mathbf{f}^{\mathrm{T}}$ as

$$\mathbf{f}^{\mathrm{T}} = [\phi(\lambda_1), \phi(\lambda_2), \ldots, \phi(\lambda_p)][C \operatorname{adj}(\lambda_1 I - A)Bk, \ldots, C \operatorname{adj}(\lambda_p I - A)Bk]^{-1} \tag{4.43}$$

for the given k, where $\phi(s) = \det(sI - A)$ and $\lambda_1, \ldots, \lambda_p$ are the poles to be assigned.

P4.6 Show that the pair $(A, \mathbf{b})$ is controllable if

$$(A, \mathbf{b}) = \left[\begin{pmatrix} 0 & 1 & 0 \\ 0 & 0 & 1 \\ -6 & -1 & 4 \end{pmatrix}, \begin{pmatrix} 0 \\ 0 \\ 1 \end{pmatrix} \right]$$

Show that the A matrix has two unstable eigenvalues and determine a state feedback control law to move these eigenvalues to -3 and -2 whilst leaving the third eigenvalue unchanged.

P4.7 Determine a state feedback control law for the system with

$$(A, \mathbf{b}) = \left[\begin{pmatrix} 0 & 1 & 0 \\ 0 & 0 & 1 \\ -1 & -2 & -2 \end{pmatrix}, \begin{pmatrix} 0 \\ 0 \\ 3 \end{pmatrix} \right]$$

to place its closed loop poles at -2, -3 and -4.

5
OPTIMAL CONTROL AND OBSERVERS

5.1 OPTIMAL CONTROL

5.1.1 Quadratic Criterion Function

Control can be considered to be a manipulation to achieve or to fulfil a given objective, such as to transfer the state to a desirable one and/or to exclude the effect of a disturbance. When more than a single unique control can fulfil the given objective, the most desirable control in the sense of minimizing a given criterion function can be used. Such a control is said to be an optimal control. If the criterion function is the time required to attain the objective state, the minimum time control is optimal, and if the criterion function is the energy required, then the minimum energy control becomes optimal.

Many kinds of functionals can be considered as criterion functions although in this book, only the quadratic criterion function is discussed since it is mathematically tractable and thus commonly used for the design of controls for linear multivariable systems.

For the linear controllable multivariable system

$$\dot{\mathbf{x}} = A\mathbf{x} + B\mathbf{u}, \mathbf{x}(0) = \mathbf{x}_0 \tag{5.1}$$

with controllable pair (A, B), we consider the following quadratic criterion function

$$J = \frac{1}{2}\int_0^\infty (\|\mathbf{x}\|_Q^2 + \|\mathbf{u}\|_R^2)\,\mathrm{d}t \qquad (5.2)^\dagger$$

where R is a symmetric positive definite matrix, and Q is a symmetric positive semidefinite matrix written as $Q = H^\mathrm{T}H$ with observable pair (A, H). The optimal control problem for a linear multivariable system with the quadratic criterion function is one of the most common problems in linear system theory. The optimal control for the quadratic function is given by the following theorem.

† $\|\mathbf{x}\|_Q^2$ denotes the quadratic form $\mathbf{x}^\mathrm{T}Q\mathbf{x}$; i.e., $\|\mathbf{x}\|_Q^2 = \mathbf{x}^\mathrm{T}Q\mathbf{x}$, $\|\mathbf{u}\|_R^2 = \mathbf{u}^\mathrm{T}R\mathbf{u}$.

Theorem 5.1

The optimal control for the controllable system (5.1) which minimizes the criterion function

$$J = \frac{1}{2}\| \mathbf{x}(t_f)\|^2_{P_f} + \frac{1}{2}\int_0^{t_f} (\| \mathbf{x}(t)\|^2_Q + \| \mathbf{u}(t)\|^2_R)\, \mathrm{d}t \tag{5.3}$$

is given by

$$\mathbf{u}(t) = -R^{-1}B^{\mathrm{T}}P(t; P_f, t_f)\mathbf{x}(t) \tag{5.4}$$

where P is positive definite, Q is positive semidefinite written as $Q = H^{\mathrm{T}}H$ with observable pair (A, H), R is a positive definite symmetric matrix, and $P(t; P_f, t_f)$ is a solution of the Riccati differential equation

$$\dot{P} = A^{\mathrm{T}}P + PA + Q - PBR^{-1}B^{\mathrm{T}}P \tag{5.5a}$$

with the terminal condition

$$P(t_f) = P_f \tag{5.5b}$$

The minimum value of the criterion function is given by

$$\min\ J = \tfrac{1}{2}\| \mathbf{x}_0 \|^2_{P(0;\, P_f,\, t_f)} \tag{5.6}$$

The proof is given in the appendix at the end of the chapter.

In particular when $t_f \to \infty$ and $P_f \to 0$, let the limit of the solution of (5.1), if it exists, be $P(t)$, that is

$$\lim_{t_f \to \infty} P(t; 0, t_f) = P(t). \tag{5.7}$$

It can be shown that if the system (5.1) is controllable the limit exists. This can be proved as follows. When the system is controllable, there exists a control to make the state zero in a finite time t_2 $(< t_f)$, so there exists a β such that

$$\begin{aligned} \min\ J &= \tfrac{1}{2}\| \mathbf{x}_0 \|^2_{P(0;\, P_f,\, t_f)} \\ &\leqslant \tfrac{1}{2}\int_0^{t_2} (\| \mathbf{x}(t)\|^2_Q + \| \mathbf{u}(t) \|^2_R)\, \mathrm{d}t \\ &\leq \beta \| \mathbf{x}_0 \|^2 \end{aligned} \tag{5.8}$$

for finite matrices Q and R. Equation (5.8) shows that the limit of the solution $P(t)$ exists, since $P(0; P_f, t_f)$ is bounded and non-decreasing with respect to t_f. For constant matrices A, B, Q, R, the limit of the solution does not depend on a shift of time and satisfies

$$P(t) = \lim_{t_f \to \infty} P(t_f + h, 0, t_f + h) = P(t + h),\ \forall h$$

Since $P(t)$ is positive definite from (5.6) (see Appendix) when (A, H) is observable, $P(t)$ is the constant, symmetric, positive definite matrix solution of the algebraic Riccati equation

$$A^{\mathrm{T}}P + PA + Q - PBR^{-1}B^{\mathrm{T}}P = 0 \tag{5.9}$$

Corollary 5.1

If the linear system (5.1) is controllable and (A, H) is observable, the optimal control minimizing the criterion function (5.2) is given by

$$\mathbf{u}(t) = -R^{-1}B^{\mathrm{T}}P\mathbf{x}(t) \tag{5.4$'$}$$

where P is the symmetric, positive definite solution of (5.9).

Example 5.1 Find the optimal control with the criterion function

$$J = \int_0^\infty \left[\mathbf{x}^{\mathrm{T}}\begin{pmatrix} q_1 & 0 \\ 0 & q_2 \end{pmatrix}\mathbf{x} + r\|\mathbf{u}\|^2\right] \mathrm{d}t \tag{5.10}$$

for the system

$$\dot{\mathbf{x}} = \begin{pmatrix} 0 & 1 \\ -\alpha_0 & -\alpha_1 \end{pmatrix}\mathbf{x} + \begin{pmatrix} 0 \\ 1 \end{pmatrix}u, \mathbf{x}(0) = \mathbf{x}_0 \tag{5.11}$$

Let the solution of the Riccati equation be denoted by

$$P = \begin{pmatrix} p_{11} & p_{12} \\ p_{12} & p_{22} \end{pmatrix}$$

then from (5.9)

$$\begin{pmatrix} 0 & -\alpha_0 \\ 1 & -\alpha_1 \end{pmatrix}\begin{pmatrix} p_{11} & p_{12} \\ p_{12} & p_{22} \end{pmatrix} + \begin{pmatrix} p_{11} & p_{12} \\ p_{12} & p_{22} \end{pmatrix}\begin{pmatrix} 0 & 1 \\ -\alpha_0 & -\alpha_1 \end{pmatrix} + \begin{pmatrix} q_1 & 0 \\ 0 & q_2 \end{pmatrix}$$

$$- r^{-1}\begin{pmatrix} p_{11} & p_{12} \\ p_{12} & p_{22} \end{pmatrix}\begin{pmatrix} 0 \\ 1 \end{pmatrix}(0 \quad 1)\begin{pmatrix} p_{11} & p_{12} \\ p_{12} & p_{22} \end{pmatrix} = 0 \tag{5.12}$$

For the (1.1), (1.2) and (2.2) elements, respectively, we obtain

$$-2\alpha_0 p_{12} + q_1 - p_{12}^2 r^{-1} = 0$$

$$p_{11} - \alpha_1 p_{12} - \alpha_0 p_{22} - p_{12}p_{22}r^{-1} = 0$$

$$2p_{12} - 2\alpha_1 p_{22} + q_2 - p_{22}^2 r^{-1} = 0$$

which yield the solutions

$$p_{12} = -r\alpha_0 \pm r\sqrt{(\alpha_0^2 + q_1/r)}$$

$$p_{22} = -r\alpha_1 \pm r\sqrt{(\alpha_1^2 + q_2/r + 2p_{12}/r)}$$

$$p_{11} = \alpha_1 p_{12} + \alpha_0 p_{22} + p_{12}p_{22}r^{-1} = (\alpha_1 r + p_{22})(\alpha_0 r + p_{12})r^{-1} - \alpha_1\alpha_0 r$$

Since P is positive definite we require $p_{11} > 0, p_{22} > 0, p_{11}p_{22} - p_{12}^2 > 0$. Thus, the elements of P are

$$p_{11} = r\{\alpha_1^2 + q_2/r + 2[-\alpha_0 + \sqrt{(\alpha_0^2 + q_1/r)}]\}^{1/2}\sqrt{(\alpha_0^2 + q_1/r)} - \alpha_0\alpha_1 r \tag{5.13a}$$

$$p_{12} = -r\alpha_0 + r\sqrt{(\alpha_0^2 + q_1/r)} \qquad (5.13b)$$

$$p_{22} = -r\alpha_1 + r[\alpha_1^2 + q_2/r + 2(-\alpha_0 + \sqrt{\alpha_0^2 + q_1/r})]^{1/2} \qquad (5.13c)$$

and the optimal control is given by

$$u = -r^{-1}(0 \quad 1)P\mathbf{x}$$

$$= -[-\alpha_0 + \sqrt{(\alpha_0^2 + q_1/r)}, -\alpha_1 + \sqrt{(\alpha_1^2 + q_2/r + 2(-\alpha_0 + \sqrt{(\alpha_0^2 + q_1/r)})^{1/2}}]\mathbf{x} \qquad (5.14)$$

Characteristic properties of systems with optimal control and the numerical computation of the optimal control are discussed later.

5.1.2 The Stability of the Optimal Control System

The optimal control for the system (5.1) with the quadratic criterion function (5.2) has been proved to be given by

$$\mathbf{u} = F\mathbf{x} \qquad (5.15)$$

where

$$F = -R^{-1}B^{\mathrm{T}}P \qquad (5.16)$$

Substituting (5.15) into (5.1), the closed loop system is given by

$$\dot{\mathbf{x}} = (A + BF)\mathbf{x}, \mathbf{x}(0) = \mathbf{x}_0 \qquad (5.17)$$

The stability of this closed loop system is discussed in this section. Using (5.16), (5.9) can be rewritten as

$$(A + BF)^{\mathrm{T}}P + P(A + BF) = -Q + F^{\mathrm{T}}RF \qquad (5.18)$$

If Q and P are positive definite then from the Lyapunov equation (2.52) $(A + BF)$ is stable. Even if Q is symmetric positive semidefinite and expessed as

$$Q = H^{\mathrm{T}}H \qquad (5.19)$$

then the closed loop is still stable as shown by the following theorem.

Theorem 5.2

If the linear system (A, B, H) is controllable and observable, then the closed loop system using the optimal control (5.15) which minimizes the quadratic criterion function (5.2) is asymptotically stable.

Proof Let the symmetric positive definite solution of (5.9) be P, and define a functional

$$V(\mathbf{x}, t) = \tfrac{1}{2}\|\mathbf{x}\|_P^2 \tag{5.20}$$

which gives a positive value for any $\mathbf{x}$ except $\mathbf{x} = \mathbf{0}$. Let $\mathbf{x}$ be the solution of (5.17), then the derivative of $V(\mathbf{x})$ is derived from (5.18) as

$$\dot{V}(\mathbf{x}, t) = \frac{1}{2}\left\{\left(\frac{\mathrm{d}}{\mathrm{d}t}\mathbf{x}^{\mathrm{T}}\right)P\mathbf{x} + \mathbf{x}^{\mathrm{T}}P\frac{\mathrm{d}}{\mathrm{d}t}\mathbf{x}\right\}$$

$$= -\frac{1}{2}\{\|\mathbf{x}\|_Q^2 + \|\mathbf{x}\|_{F^{\mathrm{T}}RF}^2\}$$

$$= -\frac{1}{2}\{\|H\mathbf{x}\|^2 + \|R^{-1}B^{\mathrm{T}}P\mathbf{x}\|_R^2\} \leq -\gamma\|\mathbf{x}(t)\|^2 \leq 0 \tag{5.21}$$

Assume that $V(\mathbf{x}, t) = 0$ for any $\mathbf{x} \neq \mathbf{0}$, then $B^{\mathrm{T}}P\mathbf{x} = \mathbf{0}$ and $H\mathbf{x} = \mathbf{0}$ should be simultaneously satisfied. Therefore $F = 0$ and $H\mathbf{x} = He^{(A+BF)t}\mathbf{x}_0 = He^{At}\mathbf{x}_0 = \mathbf{0}$ which contradicts (A, H) is observable. Therefore

$$\dot{V}(\mathbf{x}, t) < 0 \qquad \forall\mathbf{x} \neq 0$$

and $V(\mathbf{x}, t)$ is a Lyapunov function satisfying (5.17) and (5.17) is asymptotically stable.

The poles of the closed loop system (5.17) are the roots of the characteristic equation

$$\det(sI - A - BF) = \det(sI - A + BR^{-1}B^{\mathrm{T}}P) = 0 \tag{5.22}$$

Example 5.2 Find the poles for the optimal closed loop system of example 5.1. Since F is 1×2 matrix, it can be written as $\mathbf{f}^{\mathrm{T}} = (f_0, f_1)$ and is given by

$$\mathbf{f}^{\mathrm{T}} = [f_0, f_1] = \{\alpha_0 - \sqrt{(\alpha_0^2 + q_1/r)},\ \alpha_1 - [\alpha_1^2 + q_2/r + 2(-\alpha_0 + \sqrt{(\alpha_0^2 + q_1/r)})]^{1/2}\} \tag{5.23}$$

B is a 2×1 matrix, so if it is denoted by $\mathbf{b}$ the characteristic equation of the closed loop system is

$$\det(sI - A - \mathbf{b}\mathbf{f}^{\mathrm{T}})$$

$$= \det\begin{bmatrix} s & -1 \\ \alpha_0 - f_0 & s + \alpha_1 - f_1 \end{bmatrix} = s^2 + \{\alpha_1^2 + q_2/r + 2[-\alpha_0 + \sqrt{(\alpha_0^2 + q_1/r)}]\}^{1/2}s$$
$$+ \sqrt{(\alpha_0^2 + q_1/r)} \tag{5.24}$$

and the poles are given by

$$\tfrac{1}{2}\{-[\alpha_1^2 + q_2/r - 2\alpha_0 + 2\sqrt{(\alpha_0^2 + q_1/r)}]^{1/2}$$
$$\pm [\alpha_1^2 + q_2/r - 2\alpha_0 - 2\sqrt{(\alpha_0^2 + q_1/r)}]^{1/2}\} \tag{5.25}$$

Example 5.3 For the linear system

$$\dot{\mathbf{x}} = \begin{pmatrix} 0 & 1 \\ 1 & 0 \end{pmatrix} \mathbf{x} + \begin{pmatrix} 0 \\ 1 \end{pmatrix} u \tag{5.26}$$

find how the poles of the closed loop system using the optimal control minimizing

$$J = \int_0^\infty \left(\mathbf{x}^\mathrm{T} \begin{pmatrix} 9 & 0 \\ 0 & 1 \end{pmatrix} \mathbf{x} + ru^2 \right) \mathrm{d}t \tag{5.27}$$

change according to the value of r.
Substituting $\alpha_0 = -1, \alpha_1 = 0, q_1 = 9$ and $q_2 = 1$ into (5.14) gives the optimal control

$$u = -\{1 + \sqrt{(1 + 9/r)}, [2 + 1/r + 2\sqrt{(1 + 9/r)}]^{1/2}\}\mathbf{x} \tag{5.28}$$

and the poles of the closed loop system are given from (5.25) as

$$\tfrac{1}{2}(-[1/r + 2 + 2\sqrt{(1 + 9/r)}]^{1/2} \pm [1/r + 2 - 2\sqrt{1 + 9/r)}]^{1/2} \tag{5.29}$$

The poles of the open loop system (5.26) are -1 and $+1$. However, as shown by Theorem 5.2 the poles of the closed loop system for any $r\ (> 0)$ are located in the left half of the complex plane. The location of the poles as r varies is shown in Figure 5.1.

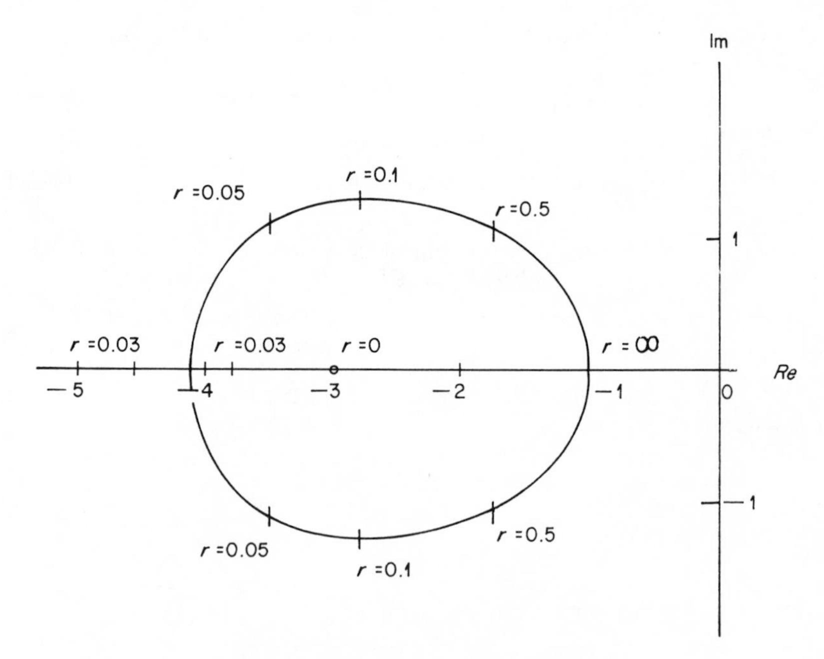

Figure 5.1 Pole locations with variation of r

5.1.3 Frequency Domain Characteristics of a System with Optimal Control

In the previous sections we have shown that the optimal control for a quadratic criterion function is given as a function of time from the solution of a Riccati equation. In this section we discuss the frequency domain properties of the single input system with optimal control.

The system considered is a controllable single input system represented by

$$\dot{\mathbf{x}} = A\mathbf{x} + \mathbf{b}u, \mathbf{x}(0) = \mathbf{x}_0 \tag{5.1$'$}$$

with the criterion function

$$J = \frac{1}{2}\int_0^{\infty} (\| \mathbf{x} \|^2_{H^{\mathrm{T}}H} + r \| u \|^2)\, \mathrm{d}t \tag{5.2$'$}$$

From Corollary 5.1, the optimal control

$$u = \mathbf{f}^{\mathrm{T}}\mathbf{x} \tag{5.15$'$}$$

is characterized by the following equations

$$\mathbf{f}^{\mathrm{T}} = -r^{-1}\mathbf{b}^{\mathrm{T}}P \tag{5.16$'$}$$

$$A^{\mathrm{T}}P + PA + H^{\mathrm{T}}H - \mathbf{f}r\mathbf{f}^{\mathrm{T}} = 0 \tag{5.18$'$}$$

This control law may be characterized in the frequency domain without using the solution of the Riccati equation by the Kalman equation.

Theorem 5.3 If (5.1)′ is controllable, the necessary and sufficient condition that the control law $\mathbf{f}^{\mathrm{T}}$ given by (5.15)′ is the optimal control minimizing (5.2)′ is that the closed loop system with feedback $\mathbf{f}$ is stable and the following Kalman equation exists for almost all ω.

$$(1 - \mathbf{f}^{\mathrm{T}}(-j\omega I - A)^{-1}\mathbf{b})(1 - \mathbf{f}^{\mathrm{T}}(j\omega I - A)^{-1}\mathbf{b}) = 1 + \frac{1}{r}\| H(j\omega I - A)^{-1}\mathbf{b} \|^2 \tag{5.30}$$

Proof Necessity. Since (5.18)′ exists for the optimal control law $\mathbf{f}^{\mathrm{T}}$, adding sP to and subtracting sP from (5.18)′ gives

$$(-sI - A)^{\mathrm{T}}P + P(sI - A) + \mathbf{f}r\mathbf{f}^{\mathrm{T}} = Q$$

Premultiplying by $(-sI - A^{\mathrm{T}})^{-1}$ and postmultiplying by $(sI - A)^{-1}$ yields

$$P(sI - A)^{-1} + (-sI - A^{\mathrm{T}})^{-1}P + r(-sI - A^{\mathrm{T}})^{-1}\mathbf{f}\mathbf{f}^{\mathrm{T}}(sI - A)^{-1}$$
$$= (-sI - A^{\mathrm{T}})^{-1}Q(sI - A)^{-1}$$

Again premultiplying by $\mathbf{b}^{\mathrm{T}}$, and postmultiplying by $\mathbf{b}$ and substituting for P from (5.16)$'$ the equation can be rewritten as

$$-r\mathbf{f}^{\mathrm{T}}(sI-A)^{-1}\mathbf{b}-r\mathbf{b}^{\mathrm{T}}(-sI-A^{\mathrm{T}})^{-1}\mathbf{f}+r\mathbf{b}^{\mathrm{T}}(-sI-A^{\mathrm{T}})^{-1}\mathbf{f}\mathbf{f}^{\mathrm{T}}(sI-A)^{-1}\mathbf{b}$$
$$=\mathbf{b}^{\mathrm{T}}(-sI-A^{\mathrm{T}})^{-1}Q(sI-A)^{-1}\mathbf{b}$$

Adding unity to each side and substituting $Q=H^{\mathrm{T}}H$ yields

$$(1-\mathbf{b}^{\mathrm{T}}(-sI-A^{\mathrm{T}})^{-1}\mathbf{f})(1-\mathbf{f}^{\mathrm{T}}(sI-A)^{-1}\mathbf{b})$$
$$=1+\frac{1}{r}\,\mathbf{b}^{\mathrm{T}}(-sI-A^{\mathrm{T}})^{-1}Q(sI-A)^{-1}\mathbf{b}$$
$$=1+\frac{1}{r}\|\,H(sI-A)^{-1}\mathbf{b}\,\|^{2} \tag{5.31}$$

which on letting $s=j\omega$ yields (5.30).

Sufficiency. Let the control law satisfying (5.16)$'$ and (5.18)$'$ be denoted by f_{*}, then from the necessary condition the control law $\mathbf{f}$ of (5.30) satisfies the equality

$$|\,1-\mathbf{f}_{*}^{\mathrm{T}}\Phi(j\omega)\mathbf{b}\,|^{2}=|\,1-\mathbf{f}^{\mathrm{T}}\Phi(j\omega)\mathbf{b}\,|^{2}$$

where

$$\Phi(s)=(sI-A)^{-1}$$

If $(A,\mathbf{b})$ is controllable, the controllable canonical form can be considered for A and $\mathbf{b}$ as

$$A=\begin{pmatrix} 0 & & \\ \vdots & & I_{n-1} \\ 0 & & \\ -a_0, & \dots & -a_{n-1}\end{pmatrix},\ \mathbf{b}=\begin{pmatrix}0\\ \vdots\\ 0\\ 1\end{pmatrix}$$

and the characteristic equation of A is

$$\phi(s)\overset{\mathrm{d}}{=}\det(sI-A)$$
$$=s^{n}+a_{n-1}s^{n-1}+\cdots+a_0$$

Let

$$\mathbf{f}^{\mathrm{T}}=(f_0,f_1,\ldots,f_{n-1})$$

then

$$\mathbf{f}^{\mathrm{T}}(j\omega\mathrm{I}-A)^{-1}\mathbf{b}=\frac{f_{n-1}s^{n-1}+\cdots+f_0}{\phi(s)}$$

So if the equality

$$\mathbf{f}^{\mathrm{T}}(j\omega I-A)^{-1}\mathbf{b}=\mathbf{f}_{*}^{\mathrm{T}}(j\omega I-A)^{-1}\mathbf{b}$$

exists, then

$$\mathbf{f} = \mathbf{f}_*$$

This means that if the decomposition of (5.30) is unique, (5.30) characterizes the optimal control law $\mathbf{f}$. The uniqueness of the decomposition of (5.30) is proved as follows: Since the numerator of the right side of (5.30) is a function of ω^2, (5.30) can be rewritten as

$$|1 - \mathbf{f}^{\mathrm{T}}(Ij\omega - A)^{-1}\mathbf{b}|^2 = 1 + \frac{1}{r}\|H(Ij\omega - A)^{-1}\mathbf{b}\|^2$$

$$= \frac{\theta(\omega^2)}{\phi(j\omega)\phi(-j\omega)}$$

Since $\theta(\omega^2)$ is positive definite, it can be decomposed into terms of the form $(\omega^2 + \alpha_{1i}^2)$, $\{(\omega + \beta_{2i})^2 + \alpha_{2i}^2\}\{(-\omega + \beta_{2i})^2 + \alpha_{2i}^2\}$ which may be further decomposed into terms $(j\omega + \alpha_{1i})$, $(-j\omega + \alpha_{1i})$, $(j\omega + \alpha_{2i} + j\beta_{2i})$, $(-j\omega + \alpha_{2i} + j\beta_{2i})$, $(j\omega + \alpha_{2i} - j\beta_{2i})$, $(-j\omega + \alpha_{2i} - j\beta_{2i})$, where α_{1i} and α_{2i} are positive. As the control law $\mathbf{f}$ makes the closed loop asymptotically stable the characteristic equation

$$\det(j\omega I - A - \mathbf{b}\mathbf{f}^{\mathrm{T}}) = \det(j\omega I - A)(1 - \mathbf{f}^{\mathrm{T}}(j\omega I - A)^{-1}\mathbf{b})$$

indicates that $1 - \mathbf{f}^{\mathrm{T}}(sI - A)^{-1}\mathbf{b} = 0$ does not have any roots in the complex right plane. From the above facts, $\theta(\omega^2)$ can be uniquely decomposed as

$$\theta(\omega^2) = p(-j\omega)p(j\omega) = |p(j\omega)|^2$$

where $p(s) = 0$ has roots in the complex left plane and $p(j\omega)$ may be written

$$p(j\omega) = \prod_i (j\omega + \alpha_{1i}) \prod_i (j\omega + \alpha_{2i} + j\beta_{2i})(j\omega + \alpha_{2i} - j\beta_{2i})$$

The Kalman equation (5.30) of Theorem 5.3 shows that when the control law stabilizing the closed loop is given, the criterion function Q for a given r can be obtained. This problem is said to be the inverse problem of optimal control.

Example 5.4 Find what criterion function is minimized by the control law

$$u = [f_0, f_1]\mathbf{x} \tag{5.32}$$

for the system of (5.11), when the control law stabilizes the closed loop system.

The characteristic equation of the closed loop system is given from (5.24) by

$$\psi_f(s) = s^2 + (\alpha_1 - f_1)s + (\alpha_0 - f_0) = 0$$

where $\alpha_1 - f_1 > 0, \alpha_0 - f_0 > 0$.

Denoting the characteristic equation of the open loop system by

$$\phi(s) = s^2 + \alpha_1 s + \alpha_0$$

the left side of (5.31) can be written as

$$1+\frac{(-f_0+f_1s)}{\phi(-s)}\frac{(-f_0-f_1s)}{\phi(s)}+\frac{-f_0-f_1s}{\phi(s)}+\frac{-f_0+f_1s}{\phi(-s)}$$

$$=1+\frac{(f_0^2-2\alpha_0f_0)-(f_1^2+2f_0-2\alpha_1f_1)s^2}{\phi(-s)\phi(s)}$$

The right side of (5.31) with $r=1$ is given by

$$1+\frac{[1,-s]Q\begin{pmatrix}1\\ s\end{pmatrix}}{\phi(-s)\phi(s)}=1+\frac{q_{11}+q_{12}s-q_{21}s-q_{22}s^2}{\phi(-s)\phi(s)}$$

where

$$Q=\begin{pmatrix}q_{11} & q_{12}\\ q_{21} & q_{22}\end{pmatrix}$$

From the equality of the above expressions, q_{12} and q_{21} which are equal, may be chosen arbitrarily and

$$\begin{aligned} q_{11}&=f_0^2-2\alpha_0f_0 & q_{22}&=f_1^2+2f_0-2\alpha_1f_1\\ &=(f_0-\alpha_0)^2-\alpha_0^2, & &=(f_1-\alpha_1)^2-\alpha_1^2+2f_0 \end{aligned}\qquad(5.33)$$

5.1.4 Square Root Locus

For Example 5.3 the poles of the closed loop system with optimal control are depicted in Figure 5.1 as r varies. These poles are roots of the Kalman equation, which is a function of the square of s as shown in (5.31), so that it has roots $\{-\lambda_i\}$ if $\{\lambda_i\}$ are roots, and the locus of these roots as a function of r is called the square root locus. Since the locus is symmetrical with respect to the imaginary axis only the left half of the figure, as shown in Figure 5.1, is usually drawn.

In this section the characteristics of the locus will be discussed using the Kalman equation.

The optimal control of (5.1)′ minimizing the criterion function (5.2)′ has been given by (5.15)′ and (5.16)′. Since there exist the relationships

$$\det(sI-A-\mathbf{b}\mathbf{f}^{\mathrm{T}})=\det\begin{bmatrix}sI-A & \mathbf{b}\\ \mathbf{f}^{\mathrm{T}} & 1\end{bmatrix}$$

$$=\det\begin{bmatrix}sI-A & 0\\ \mathbf{f}^{\mathrm{T}} & 1-\mathbf{f}^{\mathrm{T}}(sI-A)^{-1}\mathbf{b}\end{bmatrix}$$

$$=\det(sI-A)(1-\mathbf{f}^{\mathrm{T}}(sI-A)^{-1}\mathbf{b})$$

and

$$\phi(-s)\phi(s)=\det(-sI-A)\det(sI-A)$$

multiplying both sides of (5.31) by $\phi(-s)\phi(s)$ we obtain

$$\phi_f(-s)\phi_f(s)=\phi(-s)\phi(s)+\rho\phi(-s)\|H(sI-A)^{-1}\mathbf{b}\|^2\phi(s)\qquad(5.34)$$

where $\rho = 1/r$ and

$$\phi_f(s) = \det(sI - A - \mathbf{b}\mathbf{f}^{\mathrm{T}}) \tag{5.35}$$

Now, since $\| H(sI - A)^{-}\mathbf{b} \|^2$ can be expressed as

$$\| H(sI - A)^{-1}\mathbf{b} \|^2 = \frac{m(-s)m(s)}{\phi(-s)\phi(s)} \tag{5.36}$$

substituting (5.36) into (5.34) yields

$$\phi_f(-s)\phi_f(s) = \phi(-s)\phi(s) + \rho m(-s)m(s) \tag{5.34$'$}$$

Since $A, \mathbf{b}, H$ is controllable and observable then from Theorem 5.2 all the roots of $\phi_f(s) = 0$ have negative real parts, thus all the roots of the right-hand side of (5.34)$'$ with negative real parts are those of $\phi_f(s) = 0$. The roots of $\phi_f(-s)\phi_f(s) = 0$ as ρ varies constitute the square root locus.

Letting

$$\phi(s) = \prod_{i=1}^{n} (s - p_i), \; m(s) = \alpha \prod_{i=1}^{l} (s - q_i) \tag{5.37}$$

(5.34)$'$ can be rewritten as

$$\prod_{i=1}^{n} (s - p_i)(s + p_i) + \rho\alpha^2(-1)^{n-l} \prod_{i=1}^{l} (s - q_i)(s + q_i) = 0. \tag{5.38}$$

From (5.38) when $\rho = 0$ the roots of $\phi_f(s) = 0$ are those of $\phi(-s)\phi(s) = 0$ with negative real part, and when $\rho \to \infty$, l roots of $\phi_f(s) = 0$ are those of $m(-s)m(s) = 0$ with negative real part.

From (5.38) the rest of the roots are infinite zeros satisfying

$$s^{2n} + \rho\alpha^2(-1)^{n-l}s^{2l} \cong 0 \qquad (s \to \infty)$$

Letting $\rho \to \infty$, $(n - l)$ roots of $\psi_f(s) = 0$ are given by

$$s = \begin{cases} (-\rho\alpha^2)^{1/2(n-l)}: \; n - l \text{ is even} \\ (\rho\alpha^2)^{1/2(n-l)}: \; n - l \text{ is odd} \end{cases} \tag{5.39}$$

whose real parts are negative. This we can also write as

$$s = (\rho\alpha^2)^{1/2(n-1)} \, \mathrm{e}^{j\theta} \tag{5.40}$$

where

$$\theta = \frac{k\pi}{2(n-l)} \begin{cases} k = \pm 1, \pm 3, \ldots, & \text{for } n - l \text{ even} \\ k = 0, \pm 2, \ldots, & \text{for } n - l \text{ odd.} \end{cases}$$

In the case of Example 5.3,

$$\phi(-s)\phi(s) = (s^2 - 1)^2, \; m(-s)m(s) = 9 - s^2$$

so that $m(s) = s + 3$, $n = 2$ and $l = 1$, and the left half of the square root locus is as given in Figure 5.1.

5.1.5 Computational Methods of Optimal Control

To compute the optimal control for a given quadratic criterion function, the positive definite solution P of the algebraic Riccati equation (5.9) must be calculated. In the case of a single input system, however, the optimal control can be directly calculated from the Kalman equation (5.31) without the need to first obtain P. In this section computational procedures for obtaining the optimal control solution for a single input system and for the positive definite solution of the algebraic Riccati equation are described.

Computation of the optimal control of a single input system

For a single input system, $A, \mathbf{b}, Q, r$ are assumed to be given and the optimal control $\mathbf{f}^{\mathrm{T}}$ may be calculated as follows:

1st step: Multiply (5.31) by $\phi(s)\phi(-s)$ to give

$$\det(-sI - A - \mathbf{b}\mathbf{f}^{\mathrm{T}})\det(sI - A - \mathbf{b}\mathbf{f}^{\mathrm{T}})$$
$$= \phi(-s)\phi(s) + \frac{1}{r}\mathbf{b}^{\mathrm{T}}\operatorname{adj}(-sI - A^{\mathrm{T}})Q\operatorname{adj}(sI - A)\mathbf{b} \quad (5.41)$$

where

$$(sI - A)^{-1} = \frac{1}{\det(sI - A)}\operatorname{adj}(sI - A) \quad (5.42)$$

The roots of the right side of (5.41) which are of the form

$$\lambda_1, -\lambda_1, \lambda_2, -\lambda_2, \ldots, \lambda_n, -\lambda_n$$

are calculated.
2nd step: Find the n roots with negative real parts from the above roots of (5.41). Let them be denoted by $\{\lambda_i\}$, and calculate

$$\phi_f(s) \overset{d}{=} \prod_{i=1}^{n}(s - \lambda_i) = s^n + \alpha_{n-1}s^{n-1} + \cdots + \alpha_0, \ \operatorname{Re}\lambda_i < 0 \ (i = 1, \ldots, n)$$

3rd step: Find the control law $\mathbf{f}^{\mathrm{T}}$ satisfying

$$\det(sI - A - \mathbf{b}\mathbf{f}^{\mathrm{T}}) = s^n + \alpha_{n-1}s^{n-1} + \cdots + \alpha_0 \quad (5.43)$$

This can be done using the Ackermann algorithm

$$\mathbf{f}^{\mathrm{T}} = -[0, \ldots, 0 \quad 1][\mathbf{b}, A\mathbf{b}, \ldots, A^{n-1}\mathbf{b}]^{-1}\phi_f(A) \quad (5.44)$$

In the case of a multi-input system where a control to stabilize the closed loop system is to be determined one approach is to choose the first input u_1 as the optimal control of $(A, \mathbf{b}_1)$ such that

$$u_1 = \mathbf{f}_1^{\mathrm{T}}\mathbf{x} \quad (5.45)$$

and then u_2 is calculated for $(A + \mathbf{b}_1\mathbf{f}_1^\mathsf{T}, \mathbf{b}_2)$ and so on where u_i is determined for $(A + \sum_{j=1}^{i-1} \mathbf{b}_j\mathbf{f}_j^\mathsf{T}, \mathbf{b}_i)$.

Positive definite solution of the algebraic Riccati equation

The computation of the matrix solution P of the algebraic Riccati equation is quite complicated using any method. Kalman proposed the calculation of P as $P(-\infty)$ of the Riccati differential equation calculated backward from $P(t_f) = 0$. Kleinman suggested calculation of P from the limit of the iteration

$$(A - BR^{-1}B^\mathsf{T}P_i)^\mathsf{T}P_{i+1} + P_{i+1}(A - BR^{-1}B^\mathsf{T}P_i) = -Q + P_iBR^{-1}B^\mathsf{T}P_i$$

Both these methods sometimes require long computation times if the initial value is not appropriately chosen. This section describes a more straightforward computation method due to Potter.

The method proceeds as follows:
1st step: The $2n \times 2n$ matrix

$$\mathscr{A} = \begin{pmatrix} A & -BR^{-1}B^\mathsf{T} \\ -Q & -A^\mathsf{T} \end{pmatrix} \tag{5.46}$$

is calculated for (A, B, Q, R), and all its eigenvalues $\lambda_1, -\lambda_1, \lambda_2, -\lambda_2, \ldots, \lambda_n, -\lambda_n$ obtained. $\{\lambda_i\}$ and $\{-\lambda_i\}$ are the eigenvalues since

$$\begin{aligned}
&\det[sI - \mathscr{A}] \\
&= \det\begin{pmatrix} sI - A & BR^{-1}B^\mathsf{T} \\ Q & sI + A^\mathsf{T} \end{pmatrix} \\
&= \det\begin{pmatrix} sI - A + BR^{-1}B^\mathsf{T}P & BR^{-1}B^\mathsf{T} \\ Q + (sI + A^\mathsf{T})P - P(sI - A + BR^{-1}B^\mathsf{T}P) & sI + A^\mathsf{T} - PBR^{-1}B^\mathsf{T} \end{pmatrix} \\
&= \det(sI - A + BR^{-1}B^\mathsf{T}P)\det(sI + A^\mathsf{T} - PBR^{-1}B^\mathsf{T})
\end{aligned}$$

where the (2.1) block is seen from the Riccati equation to be a small block.
2nd step:
Let the eigenvector or generalized eigenvector corresponding to λ_i, which has negative real part a_i, be

$$\begin{pmatrix} \mathbf{v}_i \\ \mathbf{u}_i \end{pmatrix} \quad (i = 1, \ldots, n) \tag{5.47}$$

Then the solution of the Riccati equation is given by

$$P = [\mathbf{u}_1, \ldots, \mathbf{u}_n][\mathbf{v}_1, \ldots, \mathbf{v}_n]^{-1}. \tag{5.48}$$

The algorithm can be proved as follows: Since P satisfies

$$A^\mathsf{T}P + PA + Q - PBR^{-1}B^\mathsf{T}P = 0$$

which can be written

$$(sI + A^{\mathrm{T}})P - P(sI - A + BR^{-1}B^{\mathrm{T}}P) + Q = 0 \tag{5.49}$$

then if the eigenvalues of

$$(A - BR^{-1}B^{\mathrm{T}}P)$$

are $\lambda_1, \ldots, \lambda_n$, their real parts are negative and their corresponding eigenvectors satisfy

$$(\lambda_i I - A + BR^{-1}B^{\mathrm{T}}P)\mathbf{v}_i = 0.$$

Using this relationship in (5.49) yields

$$(\lambda_i I + A^{\mathrm{T}})P\mathbf{v}_i + Q\mathbf{v}_i = 0 \tag{5.50}$$

Letting

$$P\mathbf{v}_i = \mathbf{u}_i$$

the above two expressions may be written as

$$\begin{pmatrix} A & -BR^{-1}B^{\mathrm{T}} \\ -Q & -A^{\mathrm{T}} \end{pmatrix}\begin{pmatrix} \mathbf{v}_i \\ \mathbf{u}_i \end{pmatrix} = \lambda_i \begin{pmatrix} \mathbf{v}_i \\ \mathbf{u}_i \end{pmatrix}$$

which means that $[\mathbf{v}_i^{\mathrm{T}}, \mathbf{u}_i^{\mathrm{T}}]$ are eigenvectors corresponding to λ_i. In the case of multiple eigenvalues a similar proof can be done using generalized eigenvectors.

5.2 OBSERVERS

5.2.1 State Observer

This section is concerned with the linear system

$$\dot{\mathbf{x}} = A\mathbf{x} + B\mathbf{u}, \mathbf{x}(0) = \mathbf{x}_0 \tag{5.51a}$$

$$\mathbf{y} = C\mathbf{x} \tag{5.51b}$$

where input $\mathbf{u} \in R^m$, state $\mathbf{x} \in R^n$, output $\mathbf{y} \in R^p$, $A \in R^{n \times n}$, $B \in R^{n \times m}$, and $C \in R^{p \times n}$. It is assumed that the input and output of the system can be measured but not the state which is needed to implement the control law. This section discusses the realization of a state observer which constructs the state from the measured input and output. Before presenting the details of the observer we consider a simple problem.

Example 5.5 When the initial state $\mathbf{x}_0$ of (5.51) is unknown, can the state $\mathbf{x}$ be estimated from the state $\bar{\mathbf{x}}$ of a system with the same structure and same input

$$\dot{\bar{\mathbf{x}}} = A\bar{\mathbf{x}} + B\mathbf{u}, \bar{\mathbf{x}}(0) = \mathbf{0} \tag{5.51a)'$$

This problem can be stated as to whether the state of a system can be estimated by the state of a model system when those two systems, with different initial conditions, receive the same input.

Subtracting (5.51a)′ from (5.51a) and letting

$$\mathbf{e}(t) = \mathbf{x}(t) - \bar{\mathbf{x}}(t) \tag{5.52}$$

yields

$$\dot{\mathbf{e}} = A\mathbf{e}, \mathbf{e}(0) = \mathbf{x}_0 \tag{5.53}$$

This equation shows that provided (5.53) is asymptotically stable, that is all the eigenvalues of A have negative real parts, then $\mathbf{e}(t) \to 0$ as $t \to \infty$ which means that $\bar{\mathbf{x}}(t)$ approaches $\mathbf{x}(t)$ as t increases, otherwise $\bar{\mathbf{x}}(t)$ does not approach $\mathbf{x}(t)$ as t increases.

From this example, we see that (5.51a)′ cannot be used in all cases to estimate the state of (5.51). Thus instead of using (5.51a)′, the following system which also uses the measured output, and is given by

$$\dot{\bar{\mathbf{x}}} = A\bar{\mathbf{x}} + B\mathbf{u} + K(\mathbf{y} - C\bar{\mathbf{x}}), \bar{\mathbf{x}}(0) = \mathbf{0} \tag{5.54}$$

may be considered where $K \in R^{n \times p}$.

By substracting (5.54) from (5.51a), we obtain

$$\begin{aligned} \dot{\mathbf{e}} &= A\mathbf{e} - KC\mathbf{e} \\ &= (A - KC)\mathbf{e}, \mathbf{e}(0) = \mathbf{x}_0 \end{aligned} \tag{5.55}$$

where $\mathbf{e}(t) = \mathbf{x}(t) - \bar{\mathbf{x}}(t)$. If K is chosen such that (5.55) is asymptotically stable, then $\bar{\mathbf{x}}(t)$ approaches $\mathbf{x}(t)$ as t increases. The existence of such a K can be found from the dual of Theorems 4.2 and 4.3, that is if (A, C) is detectable, there exists a K such that $\mathbf{e}(t) \to 0$ as $t \to \infty$, and if (A, C) is observable, a K can be chosen so that the characteristic equation of (5.55) is in a given form. This means that for an observable system the error of (5.55) may be made to vanish with arbitrary modes. If (A, C) is detectable and K is chosen so that (5.55) is asymptotically stable, (5.54) is called a state observer of (5.51). This kind of observer is not always practically useful since its dimension is the same as that of the system which is often too high. Its dimension can be decreased using the output and an observer whose state $\mathbf{z}$ estimates $U\mathbf{x}$ such that

$$\begin{pmatrix} \mathbf{z}(t) \\ \mathbf{y}(t) \end{pmatrix} \to \begin{pmatrix} U \\ C \end{pmatrix} \mathbf{x}(t) \text{ as } t \to \infty \tag{5.56}$$

If the row vectors of U are linearly independent of those of C and

$$\text{rank}\begin{pmatrix} U \\ C \end{pmatrix} = n,\ U \in R^{(n-p)\times n} \tag{5.57}$$

then

$$\begin{pmatrix} U \\ C \end{pmatrix}^{-1} \begin{pmatrix} \mathbf{z}(t) \\ \mathbf{y}(t) \end{pmatrix} \to \mathbf{x}(t) \text{ as } t \to \infty \tag{5.56)'$$

This equation indicates that if $\mathbf{z}(t) \to U\mathbf{x}(t)$ $(t \to \infty)$ then $\mathbf{x}(t)$ can be estimated from $\mathbf{y}$ and $\mathbf{z}$. The dimension of $\mathbf{z}$ is the same as that of $U\mathbf{x}$, and from (5.57) the dimension of $\mathbf{z}$ to estimate the state of the system is equal to $n-p$, and it may be written as

$$\dot{\mathbf{z}} = \hat{A}\mathbf{z} + \hat{B}\mathbf{y} + \hat{J}\mathbf{u},\ \mathbf{z}(0) = \mathbf{0} \tag{5.58a}$$

$$\hat{\mathbf{x}} = [\hat{C}, \hat{D}]\begin{bmatrix} \mathbf{z} \\ \mathbf{y} \end{bmatrix} \tag{5.58b}$$

where $\hat{A} \in R^{(n-p)\times(n-p)}$, $\hat{B} \in R^{(n-p)\times p}$, $\hat{J} \in R^{(n-p)\times m}$.

To find the condition that $\mathbf{z}(t) \to U\mathbf{x}(t)$ as $t \to \infty$, premultiplying (5.51) by U and subtracting (5.58a) gives

$$\dot{\xi} = \hat{A}\xi + (UB - \hat{J})\mathbf{u} + (UA - \hat{A}U - \hat{B}C)\mathbf{x} \tag{5.55)'$$

where $\xi = U\mathbf{x} - \mathbf{z}$.

It ξ approaches zero as t increases independent of the input $\mathbf{u}(t)$ and the state, then $\hat{\mathbf{x}}(t)$ approaches $\mathbf{x}(t)$ by choosing

$$[\hat{C}, \hat{D}]\begin{pmatrix} U \\ C \end{pmatrix} = I_n \tag{5.59}$$

The system (5.58) is called a minimal order state observer.

Definition 5.1

If the linear system given by (5.58) produces the output $\hat{\mathbf{x}}$ satisfying

$$\lim_{t\to\infty} [\mathbf{x}(t) - \hat{\mathbf{x}}(t)] = 0, \qquad {}^{\forall}\mathbf{u},\ {}^{\forall}\mathbf{x}_0$$

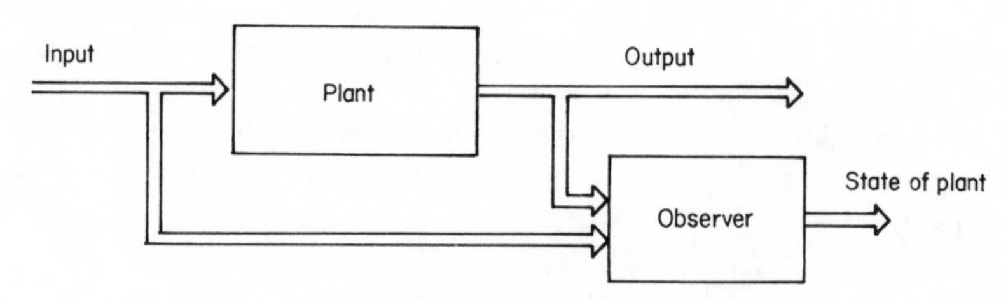

Figure 5.2 Block diagram of plant and observed

for any input and initial condition, (5.58) is a minimal order state observer of (5.51).

From (5.55)′ the condition for the observer is given by the following theorem.

Theorem 5.4

The $n-p$-dimensional observer with $\mathbf{u}$ and $\mathbf{y}$ of (5.51) as inputs, given by

$$\dot{\mathbf{z}} = \hat{A}\mathbf{z} + \hat{B}\mathbf{y} + \hat{J}\mathbf{u}, \ \mathbf{z}(0) = \mathbf{0} \tag{5.58a}$$

$$\hat{\mathbf{x}} = \hat{C}\mathbf{z} + \hat{D}\mathbf{y} \tag{5.58b}$$

is the minimal order state observer of the controllable system (5.51) if and only if it satisfies the following conditions.

(i) There exists a $U \in R^{(n-p)\times n}$ satisfying

$$UA - \hat{A}U = \hat{B}C \tag{5.60a}$$

$$\hat{J} = UB \tag{5.60b}$$

$$\hat{C}U + \hat{D}C = I_n \tag{5.60c}$$

(ii) All the eigenvalues of $\hat{A}$ have negative real parts. (5.60d)

The structure of the observer and the plant whose state is to be estimated is depicted in Figures 5.2 and 5.3.

Design methods for the minimal order state observer have been proposed by many researchers. These methods can be classified into two categories. One is to determine $(\hat{A}, \hat{B}, \hat{C}, \hat{D}, \hat{J})$ after an appropriate form of U is given.

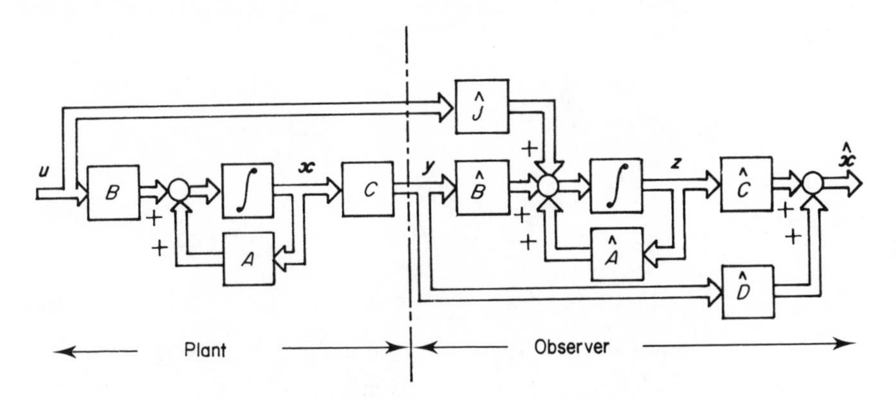

Figure 5.3 Configuration of plant and observer

The other is to determine $(\hat{A}, \hat{B})$ first, then a corresponding U is obtained to calculate $\hat{C}$, $\hat{D}$, and $\hat{J}$. A typical design algorithm for the first approach has been given by Gopinath and is presented below
First (5.60a) and (5.60c) are rewritten as

$$\begin{pmatrix} \hat{A} & \hat{B} \\ \hat{C} & \hat{D} \end{pmatrix} \begin{pmatrix} U \\ C \end{pmatrix} = \begin{pmatrix} UA \\ I_n \end{pmatrix}$$

then if $\begin{pmatrix} U \\ C \end{pmatrix}$ is non-singular, $(\hat{A}, \hat{B}, \hat{C}, \hat{D})$ can be calculated from

$$\begin{pmatrix} \hat{A} & \hat{B} \\ \hat{C} & \hat{D} \end{pmatrix} = \begin{pmatrix} UA \\ I_n \end{pmatrix} \begin{pmatrix} U \\ C \end{pmatrix}^{-1} \tag{5.61}$$

and $\hat{J}$ is obtained from (5.60b) when U is given. This minimal order state observer does not necessarily satisfy (5.60d). Gopinath proposed a design procedure using a non-singular U with prespecified structure and parameters to satisfy (5.60d).

Observer design algorithm of Gopinath

(1) The plant (A, B, C) whose state is to be estimated is considered. First $C^{\#} \in R^{(n-p)\times n}$ is constructed such that the $n \times n$ matrix

$$T^{-1} = \begin{pmatrix} C^{\#} \\ C \end{pmatrix} \in R^{n\times n} \tag{5.62}$$

is non-singular.
(2) Using T of (5.62), the equivalent system $(\bar{A}, \bar{B}, \bar{C})$ to (A, B, C) is calculated from

$$\bar{A} = T^{-1}AT = \begin{pmatrix} \bar{A}_{11} & \bar{A}_{12} \\ \bar{A}_{21} & \bar{A}_{22} \end{pmatrix} \tag{5.63a}$$

$$\bar{B} = T^{-1}B = \begin{pmatrix} \bar{B}_1 \\ \bar{B}_2 \end{pmatrix} \tag{5.63b}$$

$$\bar{C} = CT = (O, I_p) \tag{5.63c}$$

(3) For the equivalent system the minimal order state observer $(\hat{\bar{A}}, \hat{\bar{B}}, \hat{\bar{C}}, \hat{\bar{D}}, \hat{\bar{J}})$ is determined. Using $\bar{C}$ given by (5.63c). $\bar{U}$ satisfying (5.57)

is chosen as

$$\bar{U} = (I_{n-p}, \; -L) \tag{5.64}$$

(4) For $\bar{U}$ given by (5.64), the following relation is derived from (5.61).

$$\begin{pmatrix} \hat{\bar{A}} & \hat{\bar{B}} \\ \hat{\bar{C}} & \hat{\bar{D}} \end{pmatrix} = \begin{pmatrix} [I_{n-p} \quad -L]\bar{A} \\ I_n \end{pmatrix} \begin{pmatrix} I_{n-p} & -L \\ O & I_p \end{pmatrix}^{-1}$$

$$= \begin{pmatrix} \bar{A}_{11} - L\bar{A}_{21} & \bar{A}_{12} - L\bar{A}_{22} \\ I_{n-p} & O \\ O & I_p \end{pmatrix} \begin{pmatrix} I_{n-p} & L \\ O & I_p \end{pmatrix}$$

This may be written

$$\hat{\bar{A}} = \bar{A}_{11} - L\bar{A}_{21} \tag{5.65a}$$

$$\hat{\bar{B}} = \bar{A}_{11}L - L\bar{A}_{21}L + \bar{A}_{12} - L\bar{A}_{22} \tag{5.65b}$$

$$[\hat{\bar{C}}, \hat{\bar{D}}] = \begin{pmatrix} I_{n-p} & L \\ O & I_p \end{pmatrix} \tag{5.65c}$$

$$\hat{\bar{J}} = \bar{B}_1 - L\bar{B}_2 \tag{5.65d}$$

If $(\bar{A}, \bar{C})$ is observable,

$$\operatorname{rank}\begin{pmatrix} \bar{C} \\ \bar{C}\bar{A} \\ \bar{C}\bar{A}^2 \\ \vdots \\ \bar{C}\bar{A}^{n-1} \end{pmatrix} = \operatorname{rank}\begin{pmatrix} O & I_p \\ \bar{A}_{21} & \bar{A}_{22} \\ \bar{A}_{21}\bar{A}_{11} + \bar{A}_{22}\bar{A}_{21} & \bar{A}_{21}\bar{A}_{12} + \bar{A}_{22}^2 \\ \vdots & \vdots \end{pmatrix} = n \tag{5.66}$$

which indicates the first $n-p$ column vectors are linearly independent. Therefore $(\bar{A}_{21}, \bar{A}_{11})$ is observable and $\hat{\bar{A}}$ of (5.65a) can be designed to have given characteristic roots by an appropriate choice of L.

(5) Since $(\hat{\bar{A}}, \hat{\bar{B}}, \hat{\bar{C}}, \hat{\bar{D}}, \hat{\bar{J}})$ is the observer of $(\bar{A}, \bar{B}, \bar{C})$, the output of the observer $\hat{\bar{\mathbf{x}}}$ can be used to estimate the state of (A, B, C) as

$$T\hat{\bar{\mathbf{x}}} \to \mathbf{x} \text{ as } t \to \infty$$

Therefore the observer of (A, B, C) has the parameters

$$\hat{A} = \hat{\bar{A}},\ \hat{B} = \hat{\bar{B}},\ \hat{C} = T\hat{\bar{C}},\ \hat{D} = T\hat{\bar{D}},\ \hat{J} = \hat{\bar{J}}$$

Example 5.6 For the system

$$\dot{\mathbf{x}} = \begin{pmatrix} 0 & 1 & 0 \\ 0 & 0 & 1 \\ -1 & -3 & -3 \end{pmatrix}\mathbf{x} + \begin{pmatrix} 0 \\ 0 \\ 1 \end{pmatrix} u$$

$$\mathbf{y} = \begin{pmatrix} 0 & 1 & -1 \\ 1 & 0 & 0 \end{pmatrix}\mathbf{x}$$

find the observer with a single pole at -1.

Proceeding according to Gopinath's algorithm:

(1) Let

$$C^{\#} = [0 \quad 0 \quad 1]$$

then

$$T^{-1} = \begin{pmatrix} 0 & 0 & 1 \\ 0 & 1 & -1 \\ 1 & 0 & 0 \end{pmatrix} \in R^{3\times 3}$$

(2)

$$\bar{A} = T^{-1}AT = \begin{pmatrix} 0 & 0 & 1 \\ 0 & 1 & -1 \\ 1 & 0 & 0 \end{pmatrix}\begin{pmatrix} 0 & 1 & 0 \\ 0 & 0 & 1 \\ -1 & -3 & -3 \end{pmatrix}\begin{pmatrix} 0 & 0 & 1 \\ 1 & 1 & 0 \\ 1 & 0 & 0 \end{pmatrix}$$

$$= \left(\begin{array}{c|cc} -6 & -3 & -1 \\ \hline 7 & 3 & 1 \\ 1 & 1 & 0 \end{array}\right)$$

$$\bar{B} = T^{-1}B = \begin{pmatrix} 0 & 0 & 1 \\ 0 & 1 & -1 \\ 1 & 0 & 0 \end{pmatrix}\begin{pmatrix} 0 \\ 0 \\ 1 \end{pmatrix} = \begin{pmatrix} 1 \\ -1 \\ 0 \end{pmatrix}$$

$$\bar{C} = CT = \begin{pmatrix} 0 & 1 & -1 \\ 1 & 0 & 0 \end{pmatrix}\begin{pmatrix} 0 & 0 & 1 \\ 1 & 1 & 0 \\ 1 & 0 & 0 \end{pmatrix} = \begin{pmatrix} 0 & 1 & 0 \\ 0 & 0 & 1 \end{pmatrix}$$

(3) Let $\bar{U} = [1, -l_1, -l_2]$

(4) From (5.65)

$$\hat{\bar{A}} = -6 - 7l_1 - l_2$$

$$\hat{\bar{B}} = [-3 - 9l_1 - 7l_1^2 - l_1l_2 - l_2, -1 - l_1 - 7l_1l_2 - 6l_2 - l_2^2]$$

$$\hat{\bar{C}} = \begin{bmatrix} 1 \\ 0 \\ 0 \end{bmatrix}, \quad \hat{\bar{D}} = \begin{bmatrix} l_1 & l_2 \\ 1 & 0 \\ 0 & 1 \end{bmatrix}, \quad \hat{J} = 1 + l_1$$

Choosing $l_1 = -1$ and $l_2 = 2$ the observer is given by

$$\hat{\bar{A}} = -1, \quad \hat{\bar{B}} = (-1, -2)$$

$$\hat{\bar{C}} = \begin{pmatrix} 1 \\ 0 \\ 0 \end{pmatrix}, \quad \hat{\bar{D}} = \begin{pmatrix} -1 & 2 \\ 1 & 0 \\ 0 & 1 \end{pmatrix}, \quad \hat{J} = 0$$

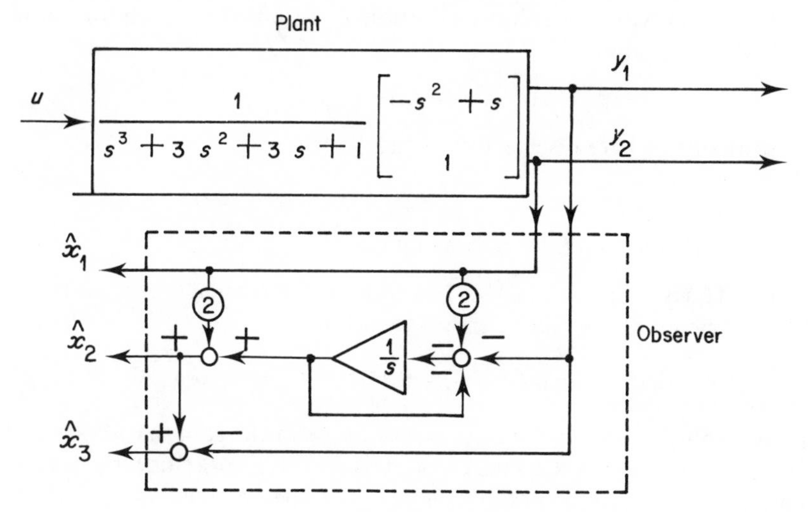

Figure 5.4 Observer for Example 5.6

(5) Therefore the observer for (A, B, C) is

$$\hat{A} = -1, \quad \hat{B} = (-1 \quad -2)$$

$$[\hat{C}, \hat{D}] = \begin{pmatrix} 0 & 0 & 1 \\ 1 & 1 & 0 \\ 1 & 0 & 0 \end{pmatrix} \left(\begin{array}{c|cc} 1 & -1 & 2 \\ 0 & 1 & 0 \\ 0 & 0 & 1 \end{array}\right) = \left(\begin{array}{c|cc} 0 & 0 & 1 \\ 1 & 0 & 2 \\ 1 & -1 & 2 \end{array}\right)$$

$$\hat{J} = 0$$

The designed observer is shown in Figure 5.4, which indicates that the observer does not use the input of the system. Therefore this observer can be used in situations where the input is unknown.

5.2.2 Determination of L in Gopinath's Algorithm

In the previous section an algorithm to design a minimal order state observer was presented. In the algorithm, one of the most important things is to determine U which is equivalent to finding L in (5.64). When L is given, the observer is designed from (5.65). L should be given so that all the eigenvalues of $\hat{A} = \bar{A}_{11} - L\bar{A}_{21}$ have negative real parts. Such an L can be determined (1) by pole assignment so that the characteristic equation of $\hat{A}$, $\det(Is - \hat{A}) = 0$ has appropriate roots or (2) by the optimal feedback $-L^{\mathrm{T}}$

for $(\bar{A}_{11}^{\mathrm{T}}, \bar{A}_{21}^{\mathrm{T}})$ with a quadratic criterion function, since such an optimal control gives a stable

$$\bar{A}_{11}^{\mathrm{T}} - \bar{A}_{21}^{\mathrm{T}} L^{\mathrm{T}}$$

These algorithms are discussed in detail.

(1) Determination of L by pole assignment

If $(\bar{A}_{11}, \bar{A}_{21})$ is observable, $(\bar{A}_{11}^{\mathrm{T}}, \bar{A}_{21}^{\mathrm{T}})$ is controllable and from Theorem 4.2, the coefficients of the characteristic equation

$$\det(sI - \bar{A}_{11} + L\bar{A}_{21}) = \det(sI - \bar{A}_{11}^{\mathrm{T}} + \bar{A}_{21}^{\mathrm{T}} L^{\mathrm{T}}) = 0$$

are known to be specified by the choice of L. In the case that $\bar{A}_{11}$ is cyclic such that there exists $\mathbf{k}$ to make $\{\mathbf{k}, \bar{A}_{11}\mathbf{k}, \ldots, \bar{A}_{11}^{n-p-1}\mathbf{k}\}$ linearly independent, an L to assign the poles of $\hat{A}$ can easily be obtained. So in this part, this design approach is given. When $\bar{A}_{11}$ is not cyclic, an L_1 to make $\bar{A}_{11} - L_1\bar{A}_{21} = \bar{A}'_{11}$ cyclic is chosen and then $\bar{A}'_{11}$ is considered as $\bar{A}_{11}$ to determine L' from

$$\bar{A}_{11} - L\bar{A}_{21} = (\bar{A}_{11} - L_1\bar{A}_{21}) - L'\bar{A}_{21}$$

When $\bar{A}_{11}$ is cyclic and $(\bar{A}_{11}, \bar{A}_{21})$ is observable, a $\boldsymbol{\gamma}^{\mathrm{T}}$ such that $(\bar{A}_{11}, \boldsymbol{\gamma}^{\mathrm{T}}\bar{A}_{21})$ is observable can be determined.†

Letting

$$L = \mathbf{l}\boldsymbol{\gamma}^{\mathrm{T}}$$

then

$$\begin{aligned}\det(sI - \hat{A}) &= \det(sI - \bar{A}_{11} + \mathbf{l}\boldsymbol{\gamma}^{\mathrm{T}}\bar{A}_{21}) \\ &= \det\begin{pmatrix} sI - \bar{A}_{11}, & \mathbf{l} \\ -\boldsymbol{\gamma}^{\mathrm{T}}\bar{A}_{21}, & 1 \end{pmatrix} \\ &= \det(sI - \bar{A}_{11})\{1 + \boldsymbol{\gamma}^{\mathrm{T}}\bar{A}_{21}(sI - \bar{A}_{11})^{-1}\mathbf{l}\} \end{aligned} \tag{5.67}$$

where

$$\det(sI - \bar{A}_{11}) = s^{n-p} + \alpha_{n-p-1}s^{n-p-1} + \cdots + \alpha_0$$

and

$$(sI - \bar{A}_{11})^{-1} = \frac{1}{\det(sI - \bar{A}_{11})}(\Gamma_{n-p-1}s^{n-p-1} + \cdots + \Gamma_0)$$

†See, for example, M. Heymann, Structure and realization problems in the theory of dynamical systems, *CISM Course and Lectures*, No. 204, Springer-Verlag (1975).

For the characteristic equation of $\hat{\bar{A}}$ to be written as

$$\det(sI - \hat{\bar{A}}) = s^{n-p} + \beta_{n-p-1}s^{n-p-1} + \cdots + \beta_0$$

(5.67) requires

$$\begin{pmatrix} \beta_{n-p-1} - \alpha_{n-p-1} \\ \beta_{n-p-2} - \alpha_{n-p-2} \\ \vdots \\ \beta_0 - \alpha_0 \end{pmatrix} = \begin{pmatrix} \boldsymbol{\gamma}^{\mathsf{T}}\bar{A}_{21}\Gamma_{n-p-1} \\ \boldsymbol{\gamma}^{\mathsf{T}}\bar{A}_{21}\Gamma_{n-p-2} \\ \vdots \\ \boldsymbol{\gamma}^{\mathsf{T}}\bar{A}_{21}\Gamma_0 \end{pmatrix} \mathbf{l}.$$

So $\mathbf{l}$ is given by

$$\mathbf{l} = \begin{pmatrix} \boldsymbol{\gamma}^{\mathsf{T}}\bar{A}_{21}\Gamma_{n-p-1} \\ \boldsymbol{\gamma}^{\mathsf{T}}\bar{A}_{21}\Gamma_{n-p-2} \\ \vdots \\ \boldsymbol{\gamma}^{\mathsf{T}}\bar{A}_{21}\Gamma_0 \end{pmatrix}^{-1} \begin{pmatrix} \beta_{n-p-1} - \alpha_{n-p-1} \\ \beta_{n-p-2} - \alpha_{n-p-2} \\ \vdots \\ \beta_0 - \alpha_0 \end{pmatrix} \tag{5.68}$$

By the dual of (4.6), it can be rewritten as

$$\mathbf{l} = \begin{pmatrix} \boldsymbol{\gamma}^{\mathsf{T}}\bar{A}_{21} \\ \boldsymbol{\gamma}^{\mathsf{T}}\bar{A}_{21}\bar{A}_{11} \\ \vdots \\ \boldsymbol{\gamma}^{\mathsf{T}}\bar{A}_{21}\bar{A}_{11}^{n-p-1} \end{pmatrix}^{-1} \begin{pmatrix} 1 & & & O \\ \alpha_{n-p-1}, 1 & & & \\ \vdots & \ddots & \ddots & \\ \alpha_1, \ldots, \alpha_{n-p-1}, & & & 1 \end{pmatrix}^{-1} \begin{pmatrix} \beta_{n-p-1} - \alpha_{n-p-1} \\ \beta_{n-p-2} - \alpha_{n-p-2} \\ \vdots \\ \beta_0 - \alpha_0 \end{pmatrix}$$

and from $\mathbf{l}\boldsymbol{\gamma}^{\mathsf{T}}$, L is determined. (5.68)′

(2) Determination of L from optimal control

The optimal feedback law $-L^{\mathsf{T}}$ for $(\bar{A}_{11}^{\mathsf{T}}, \bar{A}_{21}^{\mathsf{T}})$ with a quadratic criterion function is known to give all roots of

$$\det(sI - \bar{A}_{11} + L\bar{A}_{21}) = \det(sI - \bar{A}_{11}^{\mathsf{T}} + \bar{A}_{21}^{\mathsf{T}}L^{\mathsf{T}}) = 0$$

with negative real parts. So in this part L is determined using this approach. Since

$$\dot{\mathbf{x}} = \bar{A}_{11}^{\mathsf{T}}\mathbf{x} + \bar{A}_{21}^{\mathsf{T}}\mathbf{u}$$

is controllable, the optimal control minimizing the criterion function

$$J = \int_0^\infty \{\|\mathbf{x}\|^2_{\bar{H}^{\mathsf{T}}\bar{H}} + \|\mathbf{u}\|^2_R\}\,\mathrm{d}t$$

is given from Corollary 5.1 by

$$\mathbf{u} = -R^{-1}\bar{A}_{21}P\mathbf{x}$$

where $(\bar{A}_{11}^{\mathrm{T}}, \bar{H})$ is observable and P is the symmetrical positive definite solution of the Riccati equation

$$\bar{A}_{11}P + P\bar{A}_{11}^{\mathrm{T}} + \bar{H}^{\mathrm{T}}\bar{H} - P\bar{A}_{21}^{\mathrm{T}}R^{-1}\bar{A}_{21}P = 0 \tag{5.69}$$

From Theorem 5.2, all the eigenvalues of $\bar{A}_{11}^{\mathrm{T}} - \bar{A}_{21}^{\mathrm{T}}R^{-1}\bar{A}_{21}P$ have negative real parts and therefore

$$L^{\mathrm{T}} = R^{-1}\bar{A}_{21}P \tag{5.70}$$

makes $\hat{\bar{A}}$ of (5.65a) stable. The choice of H and R determine the poles of $\hat{\bar{A}}$.

Example 5.7 An observer is designed for the system in Example 5.6 by the method proposed in (1).
In this case

$$\bar{A}_{11} = -6$$

$$\bar{A}_{21} = \begin{pmatrix} 7 \\ 1 \end{pmatrix}$$

so $\bar{A}_{11}$ is cyclic. Let $\gamma^{\mathrm{T}} = (0 \quad 1)$

then
$$\gamma^{\mathrm{T}}\bar{A}_{21} = 1, n - p = 1$$

Since
$$\det(sI - \bar{A}_{11}) = s + 6,$$
$$\alpha_0 = 6, \Gamma_0 = 1$$

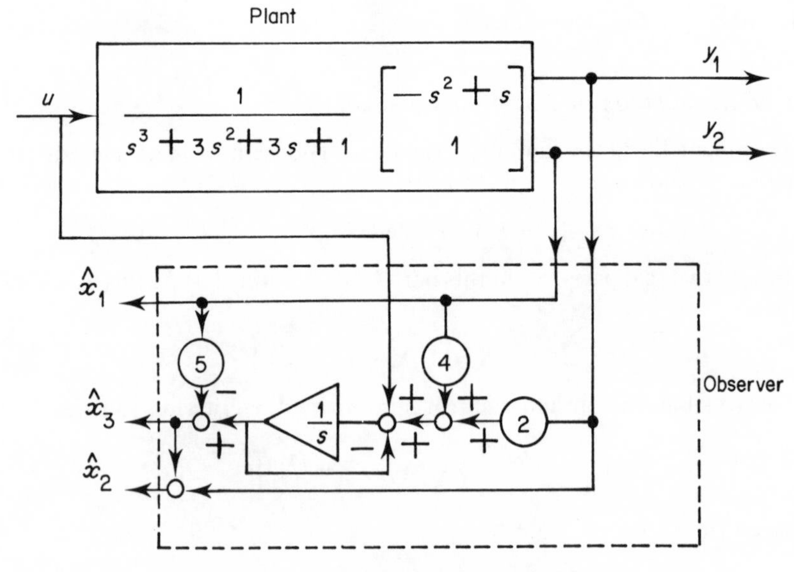

Figure 5.5 Observer for Example 5.7

and since the pole of the observer is specified as -1.

$$\det(sI - \hat{\hat{A}}) = s + 1$$

giving

$$\beta_0 = 1$$

From (5.68)

$$\mathbf{l} = (1 \cdot 1)^{-1}(1 \quad -6) = -5$$
$$L = (0, -5) \text{ that is } l_1 = 0,\ l_2 = -5$$

Thus, substituting for L in $\hat{D}$ of (4) in Example 5.6 gives

$$\hat{\hat{A}} = -1,\ \hat{\hat{B}} = [2, 4],$$

$$\hat{\hat{C}} = \begin{pmatrix} 1 \\ 0 \\ 0 \end{pmatrix},\quad \hat{\hat{D}} = \begin{pmatrix} 0 & -5 \\ 1 & 0 \\ 0 & 1 \end{pmatrix},\quad \hat{J} = 1$$

and the resulting observer following (5) of Example 5.6 is

$$\hat{A} = -1,\ \hat{B} = [2, 4]$$

$$[\hat{C}, \hat{D}] = \begin{pmatrix} 0 & 0 & 1 \\ 1 & 1 & 0 \\ 1 & 0 & 0 \end{pmatrix} \begin{pmatrix} 1 & 0 & -5 \\ 0 & 1 & 0 \\ 0 & 0 & 1 \end{pmatrix} = \begin{pmatrix} 0 & 0 & 1 \\ 1 & 1 & -5 \\ 1 & 0 & -5 \end{pmatrix}$$

$$\hat{J} = 1$$

The observer designed is shown in Figure 5.5.

From the examples, we see that the observer for the system (A, B, C) can not be uniquely determined simply by specifying the observer poles. But from (5.60) and (5.61), the observer is uniquely determined if U is specified.

5.2.3 Design of a Functional Observer

Until now all the state variables have been estimated by the observer. Generally, however, it may not be necessary to estimate all the state variables but simply $F\mathbf{x}$ which is required for control. In case F is a row vector $\mathbf{f}^{\mathrm{T}}$, the observer to estimate the scalar $\mathbf{f}^{\mathrm{T}}\mathbf{x}$ is called a functional observer and when F is a matrix, the observer to estimate $F\mathbf{x}$ is called a linear functional observer. The construction of the minimal order functional observer is a current topic, but we still do not have many powerful algorithms for its determination. In this section, a possible algorithm for designing a functional observer is presented. For the plant of (5.51), a linear functional observer to estimate $\mathbf{f}^{\mathrm{T}}\mathbf{x}$ is given by

$$\dot{\mathbf{z}} = \hat{A}\mathbf{z} + \hat{B}\mathbf{y} + \hat{J}\mathbf{u} \tag{5.58a$'$}$$

$$\hat{\mathbf{w}} = \hat{C}\mathbf{z} + \hat{D}\mathbf{y} \tag{5.58b$'$}$$

which is similar to (5.58) and it is easily shown that for a given matrix $U \in R^{q \times n}$ the following equations should be satisfied

(1) $UA - \hat{A}U = \hat{B}C$ (5.60a)′

(2) $\hat{J} = UB$ (5.60b)′

(3) $\hat{C}U + \hat{D}C = \mathbf{f}^{\mathsf{T}}$ (5.60c)′

(4) All the eigenvalues of $\hat{A}$ must be in the left half of the complex plane (5.60d)′

Comparing with (5.60), only (5.60c)′ is different and this condition is to give

$$\lim_{t \to \infty} (\hat{\mathbf{w}} - \mathbf{f}^{\mathsf{T}}\mathbf{x}) = 0 \qquad \forall \mathbf{u}, \forall \mathbf{x} \tag{5.71}$$

The minimal dimension of (5.58)′ will be shown to be at largest $\nu_0 - 1$, where ν_0 is the observability index of (A, C). Algorithms to construct a minimal functional observer with given poles have been presented by many researchers, and the following is due to A. Inoue.

Theorem 5.5

When the characteristic equation of $\hat{A}$ is given by

$$\phi(s) = s^q + \beta_{q-1}s^{q-1} + \cdots + \beta_0 \tag{5.72}$$

$\hat{A}$ can be used for a functional observer to estimate $\mathbf{f}^{\mathsf{T}}\mathbf{x}$ of (A, B, C) if and only if

$$\mathbf{f}^{\mathsf{T}}\phi(A) \in \text{row span} \begin{pmatrix} C \\ CA \\ \vdots \\ CA^q \end{pmatrix} \tag{5.73}$$

The theorem also states that the order of the functional observer q can be chosen as $\nu_0 - 1$.

Proof Necessity. If $(\hat{A}, \hat{B}, \hat{C}, \hat{D}, \hat{J})$ is a functional observer of (A, B, C), then from (5.60c)′ and (5.60a)′,

$$\begin{aligned}
\mathbf{f}^{\mathsf{T}} &= \hat{C}U + \hat{D}C \\
\mathbf{f}^{\mathsf{T}}A &= \hat{C}UA + \hat{D}CA = \hat{C}\hat{A}U + \hat{C}\hat{B}C + \hat{D}CA \\
\mathbf{f}^{\mathsf{T}}A^2 &= \hat{C}\hat{A}UA + \hat{C}\hat{B}CA + \hat{D}CA^2 = \hat{C}\hat{A}^2U + \hat{C}\hat{A}\hat{B}C + \hat{C}\hat{B}CA + \hat{D}CA^2
\end{aligned}$$

and so on.

Therefore we have

$$\begin{pmatrix} \mathbf{f}^{\mathrm{T}} \\ \mathbf{f}^{\mathrm{T}}A \\ \vdots \\ \mathbf{f}^{\mathrm{T}}A^{q} \end{pmatrix} = \begin{pmatrix} \hat{C} \\ \hat{C}\hat{A} \\ \vdots \\ \hat{C}\hat{A}^{q} \end{pmatrix} U + \begin{pmatrix} \hat{D} & 0 & \dots & 0 \\ \hat{C}\hat{B} & \hat{D} & \dots & 0 \\ \vdots & & & \\ \hat{C}\hat{A}^{q-1}\hat{B} & \hat{C}\hat{A}^{q-2}B & \dots & \hat{D} \end{pmatrix} \begin{pmatrix} C \\ CA \\ \vdots \\ CA^{q} \end{pmatrix} \tag{5.74}$$

Since $[\beta_0, \beta_1, \dots, 1] \begin{pmatrix} \hat{C} \\ \hat{C}\hat{A} \\ \vdots \\ \hat{C}\hat{A}^{q} \end{pmatrix} = \hat{C}\phi(\hat{A}) = 0$, (5.73) follows by premultiplying (5.74) by $(\beta_0, \beta_1, \dots, 1)$.

Sufficiency. This is proved by constructing a functional observer with the conditions of the theorem satisfied.

1st step: (5.73) shows that there exist $\mathbf{v}_i$ such that

$$[\mathbf{v}_0^{\mathrm{T}}, \mathbf{v}_1^{\mathrm{T}}, \dots, \mathbf{v}_q^{\mathrm{T}}] \begin{pmatrix} C \\ CA \\ \vdots \\ CA^{q} \end{pmatrix} = \mathbf{f}^{\mathrm{T}}\phi(A) \tag{5.75}$$

from which $\mathbf{v}_i$ can be determined.

2nd step: Using $\mathbf{v}_i$ obtained from the 1st step find $\hat{\mathbf{b}}_i$ from

$$\hat{\mathbf{b}}_1 = \mathbf{v}_0^{\mathrm{T}} - \beta_0\mathbf{v}_q^{\mathrm{T}} \tag{5.76a}$$

$$\hat{\mathbf{b}}_2 = \mathbf{v}_1^{\mathrm{T}} - \beta_1\mathbf{v}_q^{\mathrm{T}} \tag{5.76b}$$

$$\vdots$$

$$\hat{\mathbf{b}}_q = \mathbf{v}_{q-1}^{\mathrm{T}} - \beta_{q-1}\mathbf{v}_q^{\mathrm{T}} \tag{5.76c}$$

3rd step: Using $\hat{\mathbf{b}}_i$ and $\mathbf{v}_q^{\mathrm{T}}$ find $\mathbf{u}_i^{\mathrm{T}}$ from

$$\mathbf{u}_q^{\mathrm{T}} = \mathbf{f}^{\mathrm{T}} - \mathbf{v}_q^{\mathrm{T}}C \tag{5.77a}$$

$$\mathbf{u}_{q-1}^{\mathrm{T}} = \mathbf{u}_q^{\mathrm{T}}A + \beta_{q-1}\mathbf{u}_q^{\mathrm{T}} - \hat{\mathbf{b}}_qC \tag{5.77b}$$

$$\mathbf{u}_{q-2}^{\mathrm{T}} = \mathbf{u}_{q-1}^{\mathrm{T}}A + \beta_{q-2}\mathbf{u}_q^{\mathrm{T}} - \hat{\mathbf{b}}_{q-1}C \tag{5.77c}$$

$$\vdots$$

$$\mathbf{u}_1^{\mathrm{T}} = \mathbf{u}_1^{\mathrm{T}}A + \beta_1\mathbf{u}_q^{\mathrm{T}} - \hat{\mathbf{b}}_2C \tag{5.77d}$$

4th step: Using $\mathbf{v}_q^{\mathsf{T}}$, $\hat{\mathbf{b}}_i$ and $\mathbf{u}_i^{\mathsf{T}}$, find $(\hat{A}, \hat{B}, \hat{C}, \hat{D}, \hat{J})$ from

$$\left.\begin{aligned} &\hat{A} = \begin{pmatrix} 0 & \cdots & & -\beta_0 \\ 1 & & & \\ & \ddots & & \vdots \\ & & 1 & -\beta_{q-1} \end{pmatrix} \qquad \hat{D} = \mathbf{v}_q^{\mathsf{T}} \\ &\hat{C} = [0 \quad \ldots \quad 0 \quad 1] \\ &\hat{B} = \begin{pmatrix} \hat{\mathbf{b}}_1 \\ \vdots \\ \hat{\mathbf{b}}_q \end{pmatrix} \quad U = \begin{pmatrix} \mathbf{u}_1^{\mathsf{T}} \\ \vdots \\ \mathbf{u}_q^{\mathsf{T}} \end{pmatrix} \qquad \hat{J} = UB \end{aligned}\right\} \tag{5.78}$$

These matrices define a functional observer. Firstly since

$$\hat{C}U + \hat{D}C = \mathbf{u}_q^{\mathsf{T}} + \mathbf{v}_q^{\mathsf{T}}C = \mathbf{f}^{\mathsf{T}}$$

(5.60c)$'$ is obviously satisfied. From the 3rd step (5.77b–d) we have

$$UA - \hat{A}U = \begin{pmatrix} \mathbf{u}_1^{\mathsf{T}}A + \beta_0\mathbf{u}_q^{\mathsf{T}} \\ \mathbf{u}_2^{\mathsf{T}}A - \mathbf{u}_1^{\mathsf{T}} + \beta_0\mathbf{u}_q^{\mathsf{T}} \\ \mathbf{u}_3^{\mathsf{T}}A - \mathbf{u}_2^{\mathsf{T}} + \beta_1\mathbf{u}_q^{\mathsf{T}} \\ \vdots \\ \mathbf{u}_q^{\mathsf{T}}A - \mathbf{u}_{q-1}^{\mathsf{T}} + \beta_{q-1}\mathbf{u}_q^{\mathsf{T}} \end{pmatrix} = \begin{pmatrix} * \\ \hat{\mathbf{b}}_2^{\mathsf{T}}C \\ \vdots \\ \\ \hat{\mathbf{b}}_q C \end{pmatrix} \tag{5.79}$$

The element in the first row is found by substituting (5.77b) into (5.77c), and so on to give

$$\begin{aligned} \mathbf{u}_{q-2}^{\mathsf{T}} &= \mathbf{u}_q^{\mathsf{T}}A^2 + \beta_{q-1}\mathbf{u}_q^{\mathsf{T}}A + \beta_{q-2}\mathbf{u}_q^{\mathsf{T}} - \hat{\mathbf{b}}CA - \hat{\mathbf{b}}_{q-1}C \\ \mathbf{u}_{q-3}^{\mathsf{T}} &= \mathbf{u}_q^{\mathsf{T}}A^3 + \beta_{q-1}\mathbf{u}_q^{\mathsf{T}}A^2 + \beta_{q-2}\mathbf{u}_q^{\mathsf{T}}A + \beta_{q-3}\mathbf{u}_q^{\mathsf{T}} - \hat{\mathbf{b}}_qCA^2 - \hat{\mathbf{b}}_{q-1}CA - \hat{\mathbf{b}}_{q-2}C \\ &\vdots \\ \mathbf{u}_1^{\mathsf{T}} &= \mathbf{u}_q^{\mathsf{T}}A^{q-1} + \beta_{q-1}\mathbf{u}_q^{\mathsf{T}}A^{q-2} + \beta_{q-2}\mathbf{u}_q^{\mathsf{T}}A^{q-3} + \cdots + \beta_1\mathbf{u}_q^{\mathsf{T}} \\ &\quad - \hat{\mathbf{b}}_qCA^{q-2} - \hat{\mathbf{b}}_{q-1}CA^{q-3} - \cdots - \hat{\mathbf{b}}_2C \end{aligned}$$

Therefore, from the last expression, we can write

$$\begin{aligned} \mathbf{u}_1^{\mathsf{T}}A + \beta_0\mathbf{u}_q^{\mathsf{T}} &= \mathbf{u}_q^{\mathsf{T}}\phi(A) - [\hat{\mathbf{b}}_1, \ldots, \hat{\mathbf{b}}_q] \begin{pmatrix} C \\ CA \\ \vdots \\ CA^{q-1} \end{pmatrix} + \hat{\mathbf{b}}_1C \\ &= \mathbf{u}_q^{\mathsf{T}}\phi(A) - [\mathbf{v}_0^{\mathsf{T}}, \ldots, \mathbf{v}_{q-1}^{\mathsf{T}}] \begin{pmatrix} CA \\ CA \\ \vdots \\ CA^{q-1} \end{pmatrix} + [\beta_0, \ldots, \beta_{q-1}] \begin{pmatrix} \mathbf{v}_q^{\mathsf{T}}C \\ \mathbf{v}_q^{\mathsf{T}}CA \\ \vdots \\ \mathbf{v}_q^{\mathsf{T}}CA^{q-1} \end{pmatrix} + \hat{\mathbf{b}}_1C \end{aligned}$$

$$= (\mathbf{f}^{\mathrm{T}} - \mathbf{v}_q^{\mathrm{T}} C)\phi(A) - \mathbf{f}^{\mathrm{T}}\phi(A) + \mathbf{v}_q^{\mathrm{T}} C\phi(A) + \hat{\mathbf{b}}_1 C$$

$$= \hat{\mathbf{b}}_1 C$$

and the first row of (5.79) is $\hat{\mathbf{b}}_1 C$ as required and (5.60a)$'$ is proved. Equation (5.60b)$'$ is obvious and (5.60d)$'$ is satisfied by an appropriate choice of the characteristic equation in the design. Thus (5.78) gives a linear functional observer.

5.2.4 Characteristics of a Closed Loop System Incorporating an Observer

The optimal control law has been shown to by given be state feedback. However, in a practical situation only the input and output of a plant are often measured directly so that the state needs to be estimated using an observer.

Consider the following m-input, p-output and n-dimensional system

$$\dot{\mathbf{x}} = A\mathbf{x} + B\mathbf{u} \tag{5.51a}$$

$$\mathbf{y} = C\mathbf{x} \tag{5.51b}$$

where the feedback control law $\mathbf{u} = F\mathbf{x}$ is achieved using the estimate of the state $\hat{\mathbf{x}}$, since $\mathbf{x}$ cannot be measured directly. The estimate $\hat{\mathbf{x}}$ is given by an $n-p$ dimensional observer

$$\dot{\mathbf{z}} = \hat{A}\mathbf{z} + \hat{B}\mathbf{y} + \hat{J}\mathbf{u}, \mathbf{z}(0) = \mathbf{0} \tag{5.58a}$$

$$\hat{\mathbf{x}} = \hat{C}\mathbf{z} + \hat{D}\mathbf{y} \tag{5.58b}$$

where (5.51) and (5.58) are related with the $(n-p) \times n$ matrix U by

$$UA - \hat{A}U = \hat{B}C \tag{5.60a}$$

$$\hat{J} = UB \tag{5.60b}$$

$$\hat{C}U + \hat{D}C = I_n \tag{5.60c}$$

When the estimate $\hat{\mathbf{x}}$ is used for the control, it is given by

$$\mathbf{u} = F\hat{\mathbf{x}} \tag{5.80}$$

and the augmented system consisting of the system (5.51) and the observer (5.58) is given by

$$\begin{pmatrix} \dot{\mathbf{x}} \\ \dot{\mathbf{z}} \end{pmatrix} = \begin{pmatrix} A & O \\ \hat{B}C & \hat{A} \end{pmatrix} \begin{pmatrix} \mathbf{x} \\ \mathbf{z} \end{pmatrix} + \begin{pmatrix} B \\ \hat{J} \end{pmatrix} \mathbf{u} \tag{5.81}$$

Thus under the control law (5.80) the closed loop system is given by

$$\begin{pmatrix} \dot{\mathbf{x}} \\ \dot{\mathbf{z}} \end{pmatrix} = \begin{pmatrix} A + BF\hat{D}C & BF\hat{C} \\ \hat{B}C + \hat{J}F\hat{D}C & \hat{A} + \hat{J}F\hat{C} \end{pmatrix} \begin{pmatrix} \mathbf{x} \\ \mathbf{z} \end{pmatrix} \tag{5.81$'$}$$

and its characteristic equation is

$$\det\begin{bmatrix} sI - A - BF\hat{D}C & -BF\hat{C} \\ -\hat{B}C - \hat{J}F\hat{D}C & sI - \hat{A} - \hat{J}F\hat{C} \end{bmatrix}$$

Postmultiplying the right column of the determinant by U and adding it to the first column of the determinant gives

$$= \det\begin{bmatrix} sI - A - BF\hat{D}C - BF\hat{C}U & -BF\hat{C} \\ -\hat{B}C - \hat{J}F\hat{D}C - \hat{J}F\hat{C}U + (sI - \hat{A})U & sI - \hat{A} - \hat{J}F\hat{C} \end{bmatrix}$$

Using (5.60a) and (5.60c) the determinant can be written as

$$= \det\begin{bmatrix} sI - A - BF & -BF\hat{C} \\ -UA - \hat{J}F + sU & sI - \hat{A} - \hat{J}F\hat{C} \end{bmatrix}$$

Finally premultiplying the top row by U and subtracting it from the bottom row gives

$$= \det\begin{bmatrix} sI - A - BF & -BF\hat{C} \\ 0 & sI - \hat{A} \end{bmatrix} = \det(sI - A - BF)\det(sI - \hat{A}) \qquad (5.82)$$

which shows that the poles of the closed loop system consist of those of the observer and feedback system without the observer. This can be understood in the following way: Letting

$$\boldsymbol{\xi} = U\mathbf{x} - \mathbf{z}$$

and using the relation $\mathbf{x} - \hat{\mathbf{x}} = \hat{C}\boldsymbol{\xi}$, the control law (5.80) can be rewritten as

$$\mathbf{u} = F\mathbf{x} - F\hat{C}\boldsymbol{\xi} \qquad (5.83)$$

from (5.55)′, the closed loop system can also be written as

$$\begin{pmatrix} \dot{\mathbf{x}} \\ \dot{\boldsymbol{\xi}} \end{pmatrix} = \begin{pmatrix} A + BF & -BF\hat{C} \\ 0 & \hat{A} \end{pmatrix}\begin{pmatrix} \mathbf{x} \\ \boldsymbol{\xi} \end{pmatrix}$$

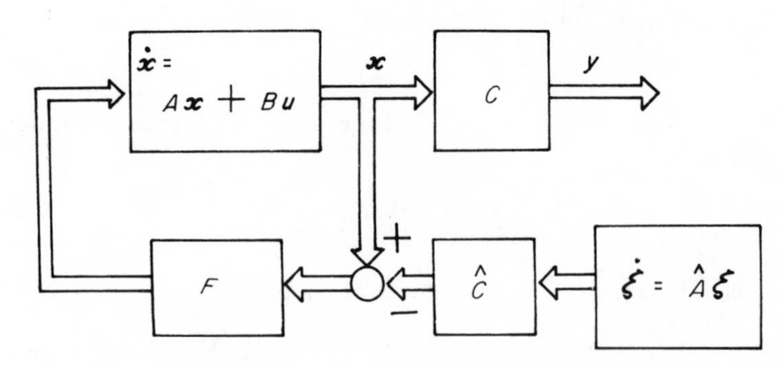

Figure 5.6 Effect of the observer on the closed loop system

This is depicted in Figure 5.6, which shows that when the observer is used, it gives a disturbance to the state feedback system due to the error produced by the observer. Thus when the control is done using $F\hat{\mathbf{x}}$ instead of the optimal control $F\mathbf{x}$ the criterion function will be degraded. This is shown by the next theorem.

Theorem 5.6

Instead of the optimal control $F\mathbf{x}$ minimizing the criterion function $J = \frac{1}{2}\int_0^\infty (\|\mathbf{x}\|_Q^2 + \|\mathbf{u}\|_R^2)dt$ the control $\mathbf{u} = F\hat{\mathbf{x}}$ using the estimate of the state from the observer deteriorates the criterion function by ΔJ where

$$\Delta J = \frac{1}{2}\|\xi(0)\|_{P_e}^2, \quad \hat{C}\xi(0) = \mathbf{x}(0) - \hat{\mathbf{x}}(0) \tag{5.84}$$

and

$$\hat{A}^{\mathrm{T}} P_e + P_e \hat{A} = -\hat{C}^{\mathrm{T}} F^{\mathrm{T}} R F \hat{C} \tag{5.85}$$

Proof From (5.3)′ in the Appendix of this chapter, the criterion function is written as $t_f \to \infty$ as

$$J = \frac{1}{2}\int_0^\infty \|\mathbf{u} - F\mathbf{x}\|_R^2 \, dt + \frac{1}{2}\|\mathbf{x}(0)\|_P^2$$

Therefore letting

$$\hat{C}\xi = \mathbf{x} - \hat{\mathbf{x}}$$

then

$$\Delta J = \frac{1}{2}\int_0^\infty \|-F\hat{C}\xi\|_R^2 \, dt$$

$$= \frac{1}{2}\|\xi(0)\|_{P_e}^2$$

where P_e is found as the solution of (5.85) from Theorem 2.11.

5.3 CONTROL SYSTEM FOR STEP COMMAND

Until now only the control required to stabilize a closed loop system has been considered. However, in the general control problem the controlled variables should be controlled to follow reference signals. A controller which makes the controlled variables follow reference signals in the presence of disturbance is called a 'servocontroller.' In the state space approach these external variables must be modelled in order to design a satisfactory control system. Here we are concerned with the design of a servo-controller.

5.3.1 Control System Design

The system to be considered is

$$\dot{\mathbf{x}} = A\mathbf{x} + B\mathbf{u} + E\mathbf{w} \tag{5.86a}$$

$$\mathbf{y} = C\mathbf{x} \tag{5.86b}$$

where the input $\mathbf{u}$ is an m-vector, the state $\mathbf{x}$ is an n-vector, the controlled vector $\mathbf{y}$ is a p-vector and the disturbance $\mathbf{w}$ is a q-vector. A, B, C, E are $n \times n$, $n \times m, p \times n, n \times q$ matrices respectively. The objective of the control is to make $\mathbf{y}$ follow the constant reference signal $\mathbf{r}$ in the presence of the constant disturbance $\mathbf{w}$ and to stabilize the closed loop. The former condition is called 'output regulation' and the latter 'internal stability'.

Let the models of the disturbance and reference signals be

$$\dot{\mathbf{x}}_d = 0, \mathbf{w} = \mathbf{x}_d \tag{5.87}$$

$$\dot{\mathbf{x}}_r = 0, \mathbf{r} = \mathbf{x}_r \tag{5.88}$$

and define the error $\mathbf{e}$ as

$$\mathbf{e} = \mathbf{y} - \mathbf{r} \tag{5.89}$$

Since $\dot{\mathbf{w}} = 0$ and $\dot{\mathbf{r}} = 0$, so the derivative of (5.86) gives

$$\ddot{\mathbf{x}} = A\dot{\mathbf{x}} + B\dot{\mathbf{u}} \tag{5.86$'$}$$

$$\dot{\mathbf{e}} = C\dot{\mathbf{x}} \tag{5.89$'$}$$

and these equations constitute an augmented system

$$\begin{pmatrix} \ddot{\mathbf{x}} \\ \dot{\mathbf{e}} \end{pmatrix} = \begin{pmatrix} A & \mathrm{O} \\ C & \mathrm{O} \end{pmatrix} \begin{pmatrix} \dot{\mathbf{x}} \\ \mathbf{e} \end{pmatrix} + \begin{pmatrix} B \\ \mathrm{O} \end{pmatrix} \dot{\mathbf{u}} \tag{5.90}$$

When $\mathbf{e}$ is taken as the output of this system, the observability matrix of this system is

$$\mathcal{O} = \begin{pmatrix} 0 & I \\ C & 0 \\ CA & 0 \\ \vdots & \vdots \end{pmatrix} \tag{5.91}$$

and it is observable if (A, C) is observable. So in order that $\mathbf{e} \to 0$ as $t \to \infty$ all the state variable $[\dot{\mathbf{x}}^{\mathrm{T}}, \mathbf{e}^{\mathrm{T}}]$ should approach zero. Since the controllability matrix of the augumented system is

$$\mathscr{C} = \begin{pmatrix} B & AB & A^2B & \dots \\ 0 & CB & CAB & \dots \end{pmatrix}$$

$$= \begin{pmatrix} A & B \\ C & 0 \end{pmatrix} \begin{pmatrix} 0 & B & AB & \dots \\ I & 0 & 0 & \dots \end{pmatrix} \tag{5.92}$$

the system is controllable if and only if

$$\operatorname{rank}\begin{pmatrix} A & B \\ C & 0 \end{pmatrix} = n + p \tag{5.93}$$

When the above condition is satisfied, the poles of the closed loop system can be assigned arbitrarily by state feedback. Such control to stabilize (5.90) is given by

$$\dot{\mathbf{u}} = F_1\dot{\mathbf{x}} + F_2\mathbf{e} \tag{5.94}$$

and $\mathbf{u}$ is given by

$$\mathbf{u}(t) = F_1\mathbf{x}(t) + F_2\int^t \mathbf{e}\,\mathrm{d}\tau + \text{constant} \tag{5.95}$$

When the constant term is taken as zero, the control law is

$$\mathbf{u} = F_1\mathbf{x} + F_2\int^t (\mathbf{y} - \mathbf{r})\,\mathrm{d}\tau \tag{5.95$'$}$$

and the closed loop is as shown in Figure 5.7, where the control system is said to have integral control.

The constant term of the control is $\mathbf{u}(0) - F_1\mathbf{x}(0)$ if the integral is taken from $\tau = 0$. Usually $\mathbf{x}(0) = \mathbf{0}$ and $\mathbf{u}(0)$ is taken as the steady state value $\mathbf{u}_s$ for the steady state of (5.86) with reference $\mathbf{r}$ and zero disturbance, that is

$$\begin{pmatrix} A & B \\ C & 0 \end{pmatrix}\begin{pmatrix} \mathbf{x}_s \\ \mathbf{u}_s \end{pmatrix} = \begin{pmatrix} 0 \\ \mathbf{r} \end{pmatrix}$$

$$\mathbf{u}_s = [0 \quad I]\begin{pmatrix} A & B \\ C & 0 \end{pmatrix}^{-1}\begin{pmatrix} 0 \\ \mathbf{r} \end{pmatrix} \overset{d}{=} N\mathbf{r} \tag{5.96}$$

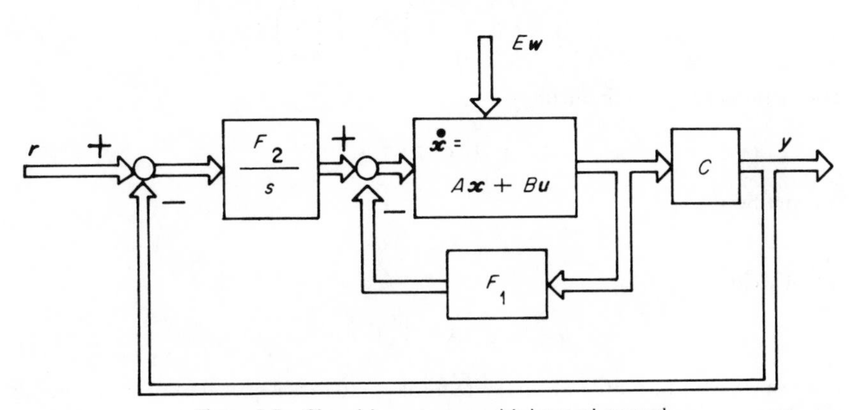

Figure 5.7 Closed loop system with integral control

And the control law is given by

$$\mathbf{u}(t) = F_1\mathbf{x}(t) + F_2 \int_0^t \mathbf{e}\,\mathrm{d}\tau + N\mathbf{r} \tag{5.97}$$

The determination of F_1 and F_2 can be done using optimal control for the criterion function

$$J = \int_0^\infty (\|\mathbf{e}\|_Q^2 + \|\dot{\mathbf{u}}\|_R^2)\,\mathrm{dt} \tag{5.98}$$

which yields

$$(F_1, F_2) = -R^{-1}(B^\mathrm{T}, 0^\mathrm{T})P \tag{5.99}$$

where P is the positive definite solution of

$$\begin{pmatrix} A & 0 \\ C & 0 \end{pmatrix}^\mathrm{T} P + P\begin{pmatrix} A & 0 \\ C & 0 \end{pmatrix} + \begin{pmatrix} 0 \\ I \end{pmatrix} Q[0 \quad I] - P\begin{pmatrix} B \\ 0 \end{pmatrix} R^{-1}[B^\mathrm{T}, 0^\mathrm{T}]P = 0 \tag{5.100}$$

Example 5.8 Design an integral controller for the constant reference signal r for the system given by

$$\dot{\mathbf{x}} = \begin{pmatrix} 0 & 1 \\ 0 & 0 \end{pmatrix}\mathbf{x} + \begin{pmatrix} 0 \\ 1 \end{pmatrix} u + \begin{pmatrix} 1 \\ 1 \end{pmatrix} w, \mathbf{x}(0) = \mathbf{0}$$

$$y = [1 \quad 0]\mathbf{x}$$

Let

$$e = y - r$$

then

$$\dot{e} = C\dot{\mathbf{x}}$$

and the augmented system is

$$\begin{pmatrix} \ddot{\mathbf{x}} \\ \dot{e} \end{pmatrix} = \begin{pmatrix} 0 & 1 & 0 \\ 0 & 0 & 0 \\ 1 & 0 & 0 \end{pmatrix}\begin{pmatrix} \dot{\mathbf{x}} \\ e \end{pmatrix} + \begin{pmatrix} 0 \\ 1 \\ 0 \end{pmatrix}\dot{u}$$

The optimal control minimizing

$$J = \int_0^\infty (\|e\|^2 + \|\dot{u}\|^2)\,\mathrm{d}t$$

is given by

$$\dot{u} = -[2, 2]\dot{\mathbf{x}} - e$$

which yields

$$u = -[2, 2]\mathbf{x} + \int^t (r - y)\,\mathrm{d}\tau$$

The response of the output of this system for $w = 0$ to a step change of the reference r is shown in Figure 5.8.

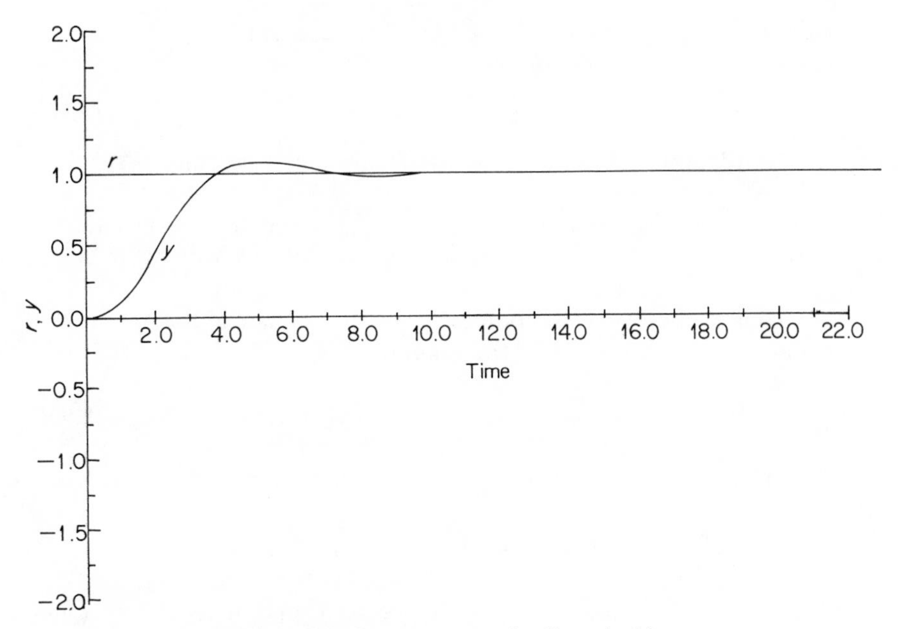

Figure 5.8 System response for Example 5.8

5.3.2 Behaviour of Observer to Constant Disturbance

In the previous section, a control law using state feedback to accommodate constant reference and disturbance signals was shown to be given by (5.95). When the state cannot be accessed directly, it may be estimated using an observer. However, when the observer is designed based on the system model without taking the disturbance into consideration the estimate given by the observer contains errors. For the system (5.86) with $\mathbf{w} = 0$, the observer is given by (5.58), that is

$$\dot{\mathbf{z}} = \hat{A}\mathbf{z} + \hat{B}\mathbf{y} + \hat{J}\mathbf{u} \tag{5.58a}$$

$$\hat{\mathbf{x}} = \hat{C}\mathbf{z} + \hat{D}\mathbf{y} \tag{5.58b}$$

with the condition

$$UA - \hat{A}U = \hat{B}C \tag{5.60a}$$

$$\hat{J} = UB \tag{5.60b}$$

$$\hat{C}U + \hat{D}C = I \tag{5.60c}$$

And $\xi = U\mathbf{x} - \mathbf{z}$ has been shown by (5.55)′ to vanish as $t \to \infty$. In the presence of the disturbance $\mathbf{w}$ in (5.86), the relationship derived in (5.55)$'$, under the condition of (5.60) becomes

$$\dot{\xi} = \hat{A}\xi + UE\mathbf{w} \tag{5.101}$$

From (5.101), provided $UE \neq 0$ and $\mathbf{w}$ is constant, ξ becomes $-\hat{A}^{-1}UE\mathbf{w}$ as $t \to \infty$.

If we use this observer in the closed loop for the control law given in the previous section then the controlled variable will coincide with the reference signal as $t \to \infty$ as shown below.

The control law (5.95)$'$ is modified, when the estimate of the state $\hat{\mathbf{x}}$ from the observer is used instead of the state, to

$$\mathbf{u} = -K_1\mathbf{x} - K_2\int^t (\mathbf{y} - \mathbf{r})\,\mathrm{d}t + K_1\hat{C}\xi \tag{5.95$''$}$$

Using the above control law for the system (5.86), the closed loop system is given by

$$\dot{\mathbf{x}} = A\mathbf{x} - BK_1\mathbf{x} + BK_2\int^t (\mathbf{r} - \mathbf{y})\,\mathrm{d}t + E\mathbf{w} + BK_1\hat{C}\xi \tag{5.102}$$

which is shown in Figure 5.9 and is interpreted as the control of the system with the disturbance $E\mathbf{w} + BK_1\hat{C}\xi$ instead of $E\mathbf{w}$. From the previous discussion of (5.101) ξ can be considered constant and given by

$$\xi = -\hat{A}^{-1}UE\mathbf{w} \tag{5.103}$$

since the observer should be designed so that the response of the observer is faster than that of the system. Since the control law (5.95)$'$ is also valid for the constant disturbance

$$E\mathbf{w} - BK_1\hat{C}\hat{A}^{-1}UE\mathbf{w}$$

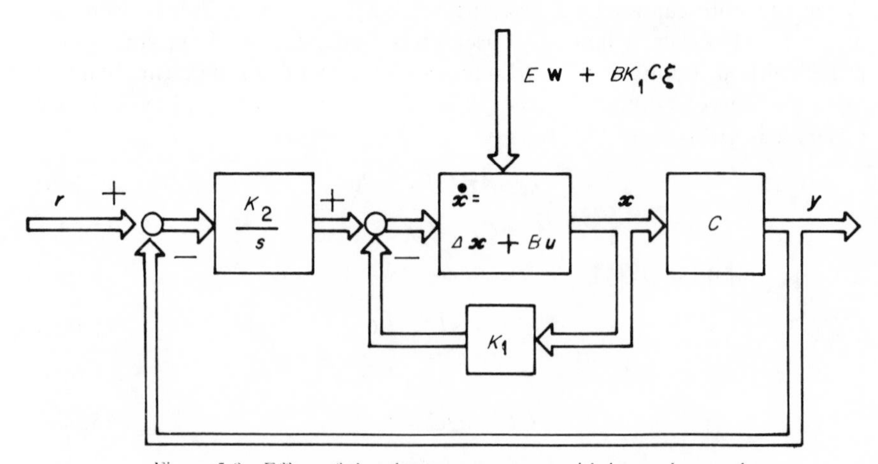

Figure 5.9 Effect of the observer on system with integral control

the controlled variable $\mathbf{y}$ approaches $\mathbf{r}$ as $t \to \infty$ even when the state in the control law is the estimate from the observer.

Example 5.9

The response of the controlled variable is considered when an observer is employed in the case of Example 5.8.

The observer is designed using the Gopinath algorithm. So it is first transformed into

$$\dot{\bar{\mathbf{x}}} = \begin{pmatrix} 0 & 0 \\ 1 & 0 \end{pmatrix} \bar{\mathbf{x}} + \begin{pmatrix} 1 \\ 0 \end{pmatrix} u$$

$$y = (0 \quad 1)\bar{\mathbf{x}}$$

using

$$T = \begin{pmatrix} 0 & 1 \\ 1 & 0 \end{pmatrix}$$

Let

$$U = (1, -l)$$

then from (5.65), the observer is given by

$$\hat{\bar{A}} = -l, \ \hat{\bar{B}} = -l^2, \ \hat{\bar{J}} = 1, \ \hat{\bar{C}} = \begin{bmatrix} 1 \\ 0 \end{bmatrix}, \ \hat{\bar{D}} = \begin{bmatrix} l \\ 1 \end{bmatrix}$$

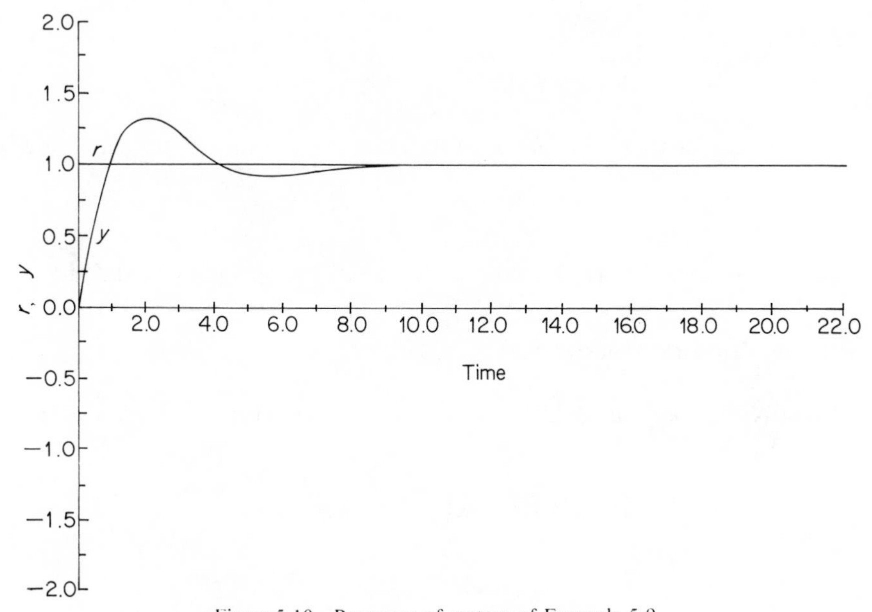

Figure 5.10 Response of system of Example 5.9

Thus using $\mathbf{x} = T\bar{\mathbf{x}}$ it follows that

$$\hat{A} = -l,\ \hat{B} = -l^2,\ \hat{J} = 1,\ \hat{C} = \begin{pmatrix} 0 \\ 1 \end{pmatrix},\quad \hat{D} = \begin{pmatrix} 1 \\ l \end{pmatrix}$$

When $l = 2$, $r = 1$ and $w = 1$ the response of the controlled variable to the step reference signal is shown in Figure 5.10.

APPENDIX

(A) Proof that $P(t)$ in Positive Definite

With the optimal control (5.4), the system is represented by

$$\dot{\mathbf{x}} = (A - BR^{-1}B^{\mathrm{T}}P)\mathbf{x}. \qquad \mathbf{x}(0) = \mathbf{x}_0$$

and the criterion function is given by

$$J = \frac{1}{2}\int_0^\infty \{\| \mathrm{e}^{(A - BR^{-1}B^{\mathrm{T}}P)t}\mathbf{x}_0 \|_Q^2 + \| R^{-1}BTP\mathrm{e}^{(A - BR^{-1}B^{\mathrm{T}}P)t}\mathbf{x}_0 \|_R^2\}\mathrm{d}t$$

Assume that (A, H) is observable and P is positive semidefinite, and let $\mathbf{x}_0$ be such that $P\mathbf{x}_0 = \mathbf{0}$, then

$$J = \frac{1}{2}\int_0^\infty \{\| \mathrm{e}^{At}\mathbf{x}_0 \|_Q^2 + \| R^{-1}B^{\mathrm{T}}P\mathrm{e}^{At}\mathbf{x}_0 \|_R^2\}\mathrm{d}t$$

$$= \frac{1}{2}\| \mathbf{x}_0 \|_P^2 = 0$$

This means that $\mathbf{x}_0$ is an element of an A invariant subspace $\mathscr{V}$ satisfying $P\mathscr{V} = 0$ and

$$H\mathrm{e}^{At}\mathbf{x}_0 = \mathbf{0}.\ \exists x_0 \in \mathscr{V}$$

which contradicts (A, H) is observable. Therefore, P is positive definite.

(B) The Proof of Theorem 5.1

If $\mathbf{x}(t)$ is the solution of (5.1), there exists for an arbitrary matrix S the relation

$$\int_0^{t_f} \frac{\mathrm{d}}{\mathrm{d}t}(\mathbf{x}^{\mathrm{T}}S\mathbf{x})\,\mathrm{d}t = \mathbf{x}^{\mathrm{T}}(t_f)S(t_f)\mathbf{x}(t_f) - \mathbf{x}^{\mathrm{T}}(0)S(0)\mathbf{x}(0)$$

$$= \int_0^{t_f}\left\{\mathbf{u}^{\mathrm{T}}B^{\mathrm{T}}S\mathbf{x} + \mathbf{x}^{\mathrm{T}}\left(A^{\mathrm{T}}S + SA + \frac{\mathrm{d}S}{\mathrm{d}t}\right)\mathbf{x} + \mathbf{x}^{\mathrm{T}}SB\mathbf{u}\right\}\mathrm{d}t$$

Since S is arbitrary, it is taken to satisfy

$$\frac{d}{dt} S + A^{T}S + SA = SBR^{-1}B^{T}S - Q, S(t_f) = P_f$$

which taking $S(t) = P(t)$ is equal to (5.5). So the following relation is derived.

$$0 = -\mathbf{x}^{T}(t_f)P_f\mathbf{x}(t_f) + \mathbf{x}^{T}(0)P(0)\mathbf{x}(0)$$
$$+ \int_{0}^{t_f} \{\mathbf{x}^{T}(PBR^{-1}B^{T}P - Q)\mathbf{x} + \mathbf{x}^{T}PB\mathbf{u} + \mathbf{u}^{T}B^{T}P\mathbf{x}\}\, dt$$

By adding the above term multiplied by 1/2 to the criterion function (5.3),

$$J = \frac{1}{2} \|\mathbf{x}(0)\|^{2}_{P(0)}$$
$$+ \frac{1}{2} \int_{0}^{t_f} \{\mathbf{x}^{T}PBR^{-1}B^{T}P\mathbf{x} + \mathbf{x}^{T}PB\mathbf{u} + \mathbf{u}^{T}B^{T}P\mathbf{x} + \mathbf{u}^{T}R\mathbf{u}\}\, dt$$
$$= \frac{1}{2} \|\mathbf{x}(0)\|^{2}_{P(0)} + \frac{1}{2} \int_{0}^{t_f} \|\mathbf{u} + R^{-1}B^{T}P\mathbf{x}\|^{2}_{R}\, dt \qquad (5.3)'$$

This equation shows that the optimal control minimizing J is given by (5.4).

(C) Proof that when (A, C) is Observable and A is Cyclic, then there Exists γ such that (A, γ^{T}C) is Observable

Such γ^{T} always exists as follows: Let (A, C) be observable, generator be $\mathbf{d}^{T}$ and $\mathcal{E}_i$ be the cyclic space generated by $\mathbf{c}_i^{T}$, then whole state space R^{n} is spanned by $\mathcal{E}_i$ as

$$R^{n} = \mathcal{E}_1 + \mathcal{E}_2 + \cdots + \mathcal{E}_p$$

and the minimal polynomial of $\mathcal{E}_i$ denotes β_i. Then the minimal polynomial of R^{n} is

$$\alpha(\lambda) = \text{LCM}(\beta_1, \ldots, \beta_m).$$

Since $\mathbf{c}_i^{T}$ is given from the cyclicity of A

$$\mathbf{c}_i^{T} = \mathbf{d}^{T}\eta_i(A), \qquad i = 1, \ldots, p$$

so if $\gamma_1, \ldots, \gamma_m$ of

$$\eta(\lambda) = \gamma_1\eta_1(\lambda) + \gamma_2\eta_2(\lambda) + \cdots + \gamma_m\eta_m(\lambda)$$

are chosen so that $\eta(\lambda)$ is co-prime to $\alpha(\lambda)$, then R^n is generated by

$$\gamma C = \sum_{i=1}^{p} \gamma_i \mathbf{c}_i^{\mathrm{T}}$$

This is using the fact that if $\mathbf{d}^{\mathrm{T}}$ is a generator of R^n with minimal polynomial $\alpha(\lambda)$, then

$$\mathbf{c}^{\mathrm{T}} = \mathbf{d}^{\mathrm{T}}\gamma(A)$$

with $\gamma(\lambda)$ which is coprime to $\alpha(\lambda)$ generates also R^n. This is because γ and α are coprime there exist ρ and σ such that

$$\gamma\rho + \alpha\sigma = 1$$

Therefore

$$\begin{aligned} \mathbf{d}^{\mathrm{T}} &= \mathbf{d}^{\mathrm{T}}\gamma(A)\rho(A) + \mathbf{d}^{\mathrm{T}}\alpha(A)\sigma(A) \\ &= \mathbf{d}^{\mathrm{T}}\gamma(A)\rho(A) \\ &= \mathbf{c}^{\mathrm{T}}\rho(A) \end{aligned}$$

This shows

$$\begin{aligned} &{}^{\forall}\mathbf{x}^{\mathrm{T}} \in R^n \\ &\mathbf{x}^{\mathrm{T}} = \mathbf{d}^{\mathrm{T}}\theta(A) = \mathbf{c}^{\mathrm{T}}\rho(A)\theta(A) \end{aligned}$$

Therefore for the minimal polynomial of $\mathbf{c}^{\mathrm{T}}$ denoted by $\beta(\lambda)$

$$\mathbf{x}^{\mathrm{T}}\beta(A) = 0 \quad \text{and} \quad \alpha \mid \beta$$

On the other hand, α is the minimal polynomial of R^n and β is that of $\mathbf{c}^{\mathrm{T}}$, so $\beta \mid \alpha$ and $\alpha = \beta$ is derived.

PROBLEMS

P5.1 Find the optimal control for a linear system

$$\dot{\mathbf{x}} = \begin{pmatrix} 0 & 1 & 0 \\ 0 & 0 & 1 \\ 0 & 0 & 0 \end{pmatrix} \mathbf{x} + \begin{pmatrix} 0 \\ 0 \\ 1 \end{pmatrix} u$$

which minimizes the criterion function

$$J = \int_0^{\infty} \left(\mathbf{x}^{\mathrm{T}} \begin{bmatrix} 3 & 0 & 0 \\ 0 & 2 & 0 \\ 0 & 0 & 1 \end{bmatrix} \mathbf{x} + u^2 \right) \mathrm{d}t$$

P5.2 When $\alpha_0 = 2$, $\alpha_1 = 3$ in the system (5.11) and $q_1 = 1$, $q_2 = 0$ in the criterion (5.10), how does the optimal control vary with γ?

P5.3 If

$$J = \frac{1}{2}\int_0^\infty (\|\mathbf{x}\|^2_Q + 2\mathbf{x}^T W\mathbf{u} + \|\mathbf{u}\|^2_R)\,dt$$

is used as the criterion function instead of (5.2), then find the optimal control $\mathbf{u}$ for (5.1).

P5.4 Design observers for the following systems.

(4–1) $$(A, \mathbf{b}, \mathbf{c}^T) = \left(\begin{bmatrix} 0 & 1 \\ -2 & -3 \end{bmatrix}, \begin{bmatrix} 0 \\ 1 \end{bmatrix}, [3 \quad 1]\right)$$

(4–2) $$(A, \mathbf{b}, \mathbf{c}^T) = \left(\begin{bmatrix} -0.5 & 0 \\ 0.2 & -0.4 \end{bmatrix}, \begin{bmatrix} 27 \\ 0 \end{bmatrix}, [0, 1]\right)$$

(4–3) $$(A, B, C) = \left(\begin{bmatrix} 0 & 0 & -1 \\ 1 & 0 & -3 \\ 0 & 1 & -3 \end{bmatrix}, \begin{bmatrix} 0 \\ 0 \\ 1 \end{bmatrix}, \begin{bmatrix} 0 & 1 & 0 \\ 0 & 0 & 1 \end{bmatrix}\right)$$

P5.5 Show that the functional observer of

$$\dot{\mathbf{x}} = \begin{bmatrix} \underbrace{\begin{matrix} 0 \\ I \end{matrix}}_{n/2} & \underbrace{\begin{matrix} A_{12} \\ A_{22} \end{matrix}}_{n/2} \end{bmatrix} \mathbf{x} + \begin{bmatrix} B_1 \\ B_2 \end{bmatrix} \mathbf{u}$$

$$\mathbf{y} = [\,\overbrace{0}\quad \overbrace{I}\,]\mathbf{x}$$

can be realized by a first order system.

6

THE KALMAN FILTER AND STOCHASTIC OPTIMAL CONTROL

6.1 THE LQG PROBLEM

We showed, in Chapter 5, that the optimal feedback control for deterministic linear systems is given in the form of a linear combination of the state variables when we wish to minimize a performance index which is quadratic in the state and control input. When all of the state variables are not accessible a state observer can play an important role in reconstructing the states to implement the optimal control.

In this chapter we focus our attention on a stochastic version of the linear optimal control problem and consider the situation where the system is perturbed by random disturbances and only some of the states can be measured through noise-corrupted output data. We confine the discussion only to the basic principles of the stochastic control problem and so some of the details of more mathematically rigorous treatments will be omitted.

The stochastic linear system considered is described by

$$\dot{\mathbf{x}}(t) = A\mathbf{x}(t) + B\mathbf{u}(t) + \Gamma\mathbf{v}(t), \qquad \mathbf{x}(0) = \mathbf{x}_0 \tag{6.1}$$

$$\mathbf{y}(t) = C\mathbf{x}(t) + \mathbf{w}(t) \tag{6.2}$$

where $\mathbf{x}(t)$ is an n-vector of the state variables, $\mathbf{u}(t)$ is an m-vector of the control input, $\mathbf{y}(t)$ is a p-vector of the measured output, $\mathbf{v}(t)$ is an r-vector of the system disturbances, and $\mathbf{w}(t)$ is a p-vector of measurement noises. The matrices A, B, Γ, and C have constant coefficients and are of dimension $n \times n$, $n \times m$, $n \times r$, and $p \times n$ respectively.

Both $\mathbf{v}(t)$ and $\mathbf{w}(t)$ are assumed to be white Gaussian random variables having zero means and to be statistically independent of each other, that is

$$E\{\mathbf{v}(t)\} = \bar{\mathbf{v}} = 0, \quad \text{cov}\{\mathbf{v}(t), \mathbf{v}(\tau)\} = V\delta(t-\tau) \tag{6.3}$$

$$E\{\mathbf{w}(t)\} = \bar{\mathbf{w}} = 0, \quad \text{cov}\{\mathbf{w}(t), \mathbf{w}(\tau)\} = W\delta(t-\tau) \tag{6.4}$$

$$\text{cov}\{\mathbf{v}(t), \mathbf{w}(t)\} = 0 \tag{6.5}$$

where $E\{\mathbf{a}\} \equiv \bar{\mathbf{a}}$ is the expected value of the random variable $\mathbf{a}$, and cov $\{\mathbf{a}, \mathbf{b}\} \equiv E\{(\mathbf{a}-\bar{\mathbf{a}})(\mathbf{b}-\bar{\mathbf{b}})^{\mathrm{T}}\}$ denotes the covariance matrix for the random variables **a** and **b**. The presence of the delta function $\delta(t-\tau)$ guarantees the whiteness property implying that the noise is uncorrelated from one instant to the next. It is also assumed that V is a non-negative definite covariance matrix and W is a positive definite covariance matrix.

The initial state $\mathbf{x}_0$ is also assumed to be a Gaussian random variable with mean $\bar{\mathbf{x}}_0$ and covariance Σ_0, and to be independent of both $\mathbf{v}(t)$ and $\mathbf{w}(t)$. Therefore, in mathematical terms,

$$E\{\mathbf{x}_0\} = \bar{\mathbf{x}}_0, \quad \operatorname{cov}\{\mathbf{x}_0, \mathbf{x}_0\} = \Sigma_0 \tag{6.6a}$$

$$E\{\mathbf{x}_0\mathbf{v}^{\mathrm{T}}(t)\} = E\{\mathbf{x}_0\mathbf{w}^{\mathrm{T}}(t)\} = 0 \tag{6.6b}$$

The control input $\mathbf{u}(t)$ should be manipulated to force the system to behave in a desired manner. A quadratic performance index is introduced, as in Chapter 5, to evaluate the quality of the system behaviour. Since $\mathbf{v}(t)$, $\mathbf{w}(t)$, and $\mathbf{x}_0$ are random variables, we need to evaluate the expected value of the performance index, that is

$$J = \tfrac{1}{2} E\left[\int_0^{t_f} \{\|\mathbf{x}(t)\|_Q^2 + \|\mathbf{u}(t)\|_R^2\}\,\mathrm{d}t\right] \tag{6.7}$$

where Q is a symmetric non-negative definite matrix and R is a symmetric positive definite matrix.

The stochastic optimization problem for the linear system described by (6.1) and (6.2) which is subjected to Gaussian random disturbances and measurement noise with a control specified to minimize the quadratic performance index (6.7) is called an LQG problem. The term LQG originates from 'Linear, Quadratic and Gaussian'. As will be shown later, the LQG problem consists of an optimal filtering and an optimal stochastic control problem. In Section 6.2, we study the Kalman filter which can estimate all of the state variables in an optimal fashion based on the set of measured output data. In Section 6.3, we then give the optimal solution for the LQG problem.

6.2 THE KALMAN FILTER

The optimal estimation problem considered in this section is that of how to estimate the state of a stochastic linear system with the assumption that the measurement data obtained is noise corrupted. The estimation operation is called optimal when an estimate is determined in accordance with the minimization of some criterion or loss function, which represents a quantitative measure of how good the estimate is. If we are concerned with the estimation error, it is reasonable to take a non-negative loss function of the estimation error such as a quadratic function.

We adopt the mean square error as the criterion, that is

$$
\begin{aligned}
I &= E\{\| \mathbf{x}(t) - \hat{\mathbf{x}}(t) \|^2\} \\
&= \operatorname{tr} E\{\tilde{\mathbf{x}}(t)\tilde{\mathbf{x}}^{\mathrm{T}}(t)\}
\end{aligned}
\tag{6.8a}
$$

where $\hat{\mathbf{x}}(t)$ is the estimate of the state $\mathbf{x}(t)$ of the linear system (6.1) which minimizes the measure (6.8a), on the assumption that a set of measurement data $\mathscr{Y}_t \equiv \{y(\tau) \text{ for } 0 \leqslant \tau \leqslant t\}$ is available and that the input function $\mathbf{u}(t)$ is known. $\tilde{\mathbf{x}}(t)$ is the estimation error defined by

$$
\tilde{\mathbf{x}}(t) \equiv \mathbf{x}(t) - \hat{\mathbf{x}}(t) \tag{6.8b}
$$

The optimal estimate minimizing the quadratic loss function (6.8a) is called the least square estimate and has the following properties:

(a) The optimal estimate $\hat{\mathbf{x}}(t)$ is the conditional expectation of $\mathbf{x}(t)$ given the measured data $\mathscr{Y}_t$, that is

$$
\hat{\mathbf{x}}(t) = \hat{\mathbf{x}}^0(t) \equiv E\{\mathbf{x}(t) \mid \mathscr{Y}_t\} \tag{6.9}
$$

(b) The estimate $\hat{\mathbf{x}}(t)$ is an unbiased estimate, which means that

$$
E\{\hat{\mathbf{x}}(t)\} = E\{\mathbf{x}(t)\} \tag{6.10}
$$

Proof Equation (6.8a) can be rewritten as

$$
I = E[E\{\| \mathbf{x}(t) - \hat{\mathbf{x}}(t) \|^2 \mid \mathscr{Y}_t\}]. \tag{6.11}
$$

Let I' be denoted by

$$
I' = E\{\| \mathbf{x}(t) - \hat{x}(t) \|^2 \mid \mathscr{Y}_t\}.
$$

Then it is noticed that the estimate $\hat{\mathbf{x}}(t)$ that minimizes the criterion I' also minimizes I. The criterion I' can be written as

$$
\begin{aligned}
I' &= E\{\| \mathbf{x}(t) - \hat{\mathbf{x}}^0(t) + \hat{\mathbf{x}}^0(t) - \hat{\mathbf{x}}(t) \|^2 \mid \mathscr{Y}_t\} \\
&= E\{\| \mathbf{x}(t) - \hat{\mathbf{x}}^0(t) \|^2 \mid \mathscr{Y}_t\} \\
&\quad + 2E\{ [\mathbf{x}(t) - \hat{\mathbf{x}}^0(t)]^{\mathrm{T}} [\hat{\mathbf{x}}^0(t) - \hat{\mathbf{x}}(t)] \mid \mathscr{Y}_t\} \\
&\quad + E\{\| \hat{\mathbf{x}}^0(t) - \hat{\mathbf{x}}(t) \|^2 \mid \mathscr{Y}_t\} \\
&= E\{\| \mathbf{x}(t) - \hat{\mathbf{x}}^0(t) \|^2 \mid \mathscr{Y}_t\} + E\{\| \hat{\mathbf{x}}^0(t) - \hat{\mathbf{x}}(t) \|^2 \mid \mathscr{Y}_t\}
\end{aligned}
$$

Therefore, $\hat{\mathbf{x}}(t)$ that minimizes I' is given by $\hat{\mathbf{x}}(t) = \hat{\mathbf{x}}^0(t)$ and (6.9) is established.

We restrict our attention to an estimate of the form

$$
\hat{\mathbf{x}}(t) = \mathbf{h}(t) + \int_0^t H(t, \tau)\mathbf{y}(\tau)\, \mathrm{d}\tau \tag{6.12}
$$

where $\mathbf{h}(t)$ is an n-vector and $H(t, \tau)$ an $n \times n$ matrix. It is assumed in (6.12) that the optimal estimate is obtained through a linear operation on

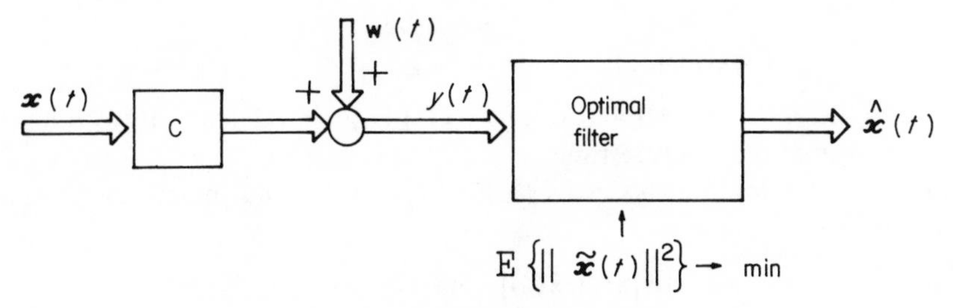

Figure 6.1 Optimal filter

the measured data observed up to the present time t. The problem is to determine both $\mathbf{h}(t)$ and $H(t, \tau)$ such that the mean square error criterion (6.8a) is minimized, as illustrated in Fig. 6.1. It is known that, if all of the random variables are Gaussian, the linear estimation given by (6.12) is optimum.

It follows from the result of the unbiased property (b) of the estimate $\hat{\mathbf{x}}(t)$ that

$$E\{\hat{\mathbf{x}}(t)\} = \mathbf{h}(t) + \int_0^t H(t, \tau)E\{\mathbf{y}(\tau)\}\, d\tau = E\{\mathbf{x}(t)\}$$

then,

$$\mathbf{h}(t) = \bar{\mathbf{x}}(t) - \int_0^t H(t, \tau)\bar{\mathbf{y}}(\tau)\, d\tau \tag{6.13}$$

where

$$\bar{\mathbf{x}}(t) \equiv E\{\mathbf{x}(t)\}, \qquad \bar{\mathbf{y}}(t) \equiv E\{\mathbf{y}(t)\} = C\bar{\mathbf{x}}(t) \tag{6.14}$$

Hence we have

$$\hat{\mathbf{x}}(t) = \bar{\mathbf{x}}(t) + \int_0^t H(t, \tau)\{\mathbf{y}(\tau) - \bar{\mathbf{y}}(\tau)\}\, d\tau \tag{6.15}$$

Now the problem is reduced to that of determining the optimal kernel $H(t, \tau)$, which is given by the following theorem.

Theorem 6.1

A necessary and sufficient condition for $\hat{\mathbf{x}}(t)$ to be the optimal estimate is the kernel $H(t, \tau)$ satisfies the Wiener–Hopf integral equation:

$$\mathrm{cov}\{\mathbf{x}(t), \mathbf{y}(\tau)\} = \int_0^t H(t, \sigma)\mathrm{cov}\{\mathbf{y}(\sigma), \mathbf{y}(\tau)\}\, d\sigma \qquad \text{for } 0 \leqslant \tau < t \tag{6.16}$$

Proof See Appendix 6(A).

Substituting (6.15) into (6.16) gives

$$\mathrm{cov}\{\mathbf{x}(t), \mathbf{y}(t)\} = \mathrm{cov}\{\hat{\mathbf{x}}(t), \mathbf{y}(\tau)\} \qquad \text{for } 0 \leqslant \tau < t$$

then

$$\text{cov}\{\tilde{\mathbf{x}}(t), \mathbf{y}(\tau)\} = 0 \qquad \text{for } 0 \leqslant \tau < t \tag{6.17}$$

This implies that the estimation error $\tilde{\mathbf{x}}(t)$ is uncorrelated with all of the past measured data before time t.

Postmultiplying both sides of (6.17) by $H^{\mathrm{T}}(t, \tau)$ and integrating over τ from 0 to t, we obtain

$$\text{cov}\{\mathbf{x}(t), \hat{\mathbf{x}}(t)\} = \text{cov}\{\hat{\mathbf{x}}(t), \hat{\mathbf{x}}(t)\}$$

or equivalently

$$\text{cov}\{\tilde{\mathbf{x}}(t), \hat{\mathbf{x}}(t)\} = 0 \tag{6.18}$$

Corollary 6.1 (orthogonal projection lemma)

The optimum estimate $\hat{\mathbf{x}}(t)$ is orthogonal to the estimation error $\tilde{\mathbf{x}}(t)$ in the sense of (6.18).

It is difficult to derive the explicit form of $H(t, \tau)$ by directly solving (6.16). Kalman approached the problem using a state variable description for the input–output structure of the optimal estimator (6.15), and derived a result known as the Kalman filter, the structure of which is illustrated in Figure 6.2.

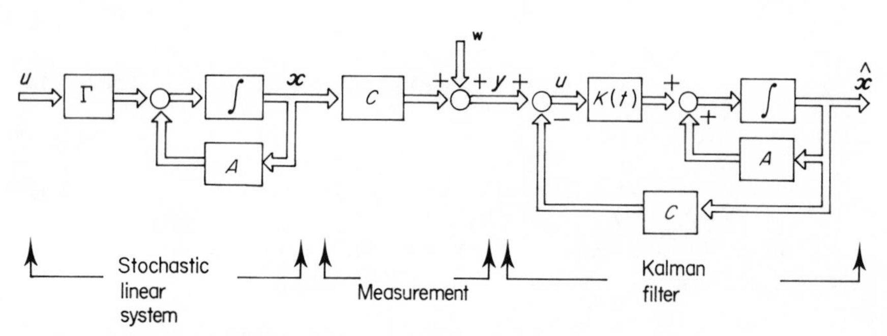

Figure 6.2 Kalman filter block diagram

Theorem 6.2 (Kalman filter)

The optimal estimate $\hat{\mathbf{x}}(t)$ for the stochastic linear system (6.1) and (6.2) subject to random noise characterized by (6.3) to (6.6) is given by

$$\dot{\hat{\mathbf{x}}}(t) = A\hat{\mathbf{x}}(t) + B\mathbf{u}(t) + K(t)[\mathbf{y}(t) - C\hat{\mathbf{x}}(t)] \qquad \hat{\mathbf{x}}(0) = \bar{\mathbf{x}}_0 \tag{6.19}$$

where

$$K(t) = \Sigma(t)C^{\mathrm{T}}W^{-1} \tag{6.20}$$

$$\dot{\Sigma}(t) = A\Sigma(t) + \Sigma(t)A^{\mathrm{T}} + \Gamma V\Gamma^{\mathrm{T}} - \Sigma(t)C^{\mathrm{T}}W^{-1}C\Sigma(t) \qquad \Sigma(0) = \Sigma_0 \quad (6.21)$$

$K(t)$ is the filter gain, and $\Sigma(t)$ is the covariance of the estimation error defined by $\Sigma(t) \equiv \mathrm{cov}\{\tilde{\mathbf{x}}(t), \tilde{\mathbf{x}}(t)\}$.

Proof See Appendix 6(B).

Example 6.1 Consider the scalar stochastic linear system

$$\dot{x}(t) = ax(t) + \gamma v(t)$$
$$y(t) = cx(t) + w(t)$$

It follows from (6.19) to (6.21) that the Kalman filter is given by

$$\dot{\hat{x}}(t) = a\hat{x}(t) + K(t)[y(t) - c\hat{x}(t)], \qquad \hat{x}(0) = \bar{x}_0$$
$$K(t) = c\Sigma(t)/W$$
$$\dot{\Sigma}(t) = 2a\Sigma(t) + \gamma^2 V - c^2\Sigma^2(t)/W, \qquad \Sigma(0) = \Sigma_0$$

The solution $\Sigma(t)$ of the above Riccati equation is given by

$$\Sigma(t) = \frac{\rho_1(\Sigma_0 - \rho_2) - \rho_2(\Sigma_0 - \rho_1)\mathrm{e}^{-2\mu t}}{\Sigma_0 - \rho_2 - (\Sigma_0 - \rho_1)\mathrm{e}^{-2\mu t}}$$

where

$$\mu = (a^2 + c^2\gamma^2 V/W)^{1/2}$$
$$\rho_1, \rho_2 = W(a \pm \mu)/c^2, \qquad \rho_1 > 0,\ \rho_2 < 0$$

As the time t tends to infinity, the solution $\Sigma(t)$ converges to the positive constant:

$$\Sigma = \lim_{t\to\infty} \Sigma(t) = W(a + \mu)/c^2$$

Then we have the stationary Kalman filter given by

$$\dot{\hat{x}}(t) = -\mu\hat{x}(t) + \frac{a+\mu}{c}\, y(t)$$

Example 6.2 Consider the following second-order system perturbed by stochastic disturbances, that is

$$\ddot{x}(t) + \omega^2 x(t) = v(t)$$
$$y(t) = x(t) + w(t).$$

Assigning the state variables as $x_1(t) = x(t)$ and $x_2(t) = \dot{x}_1(t)$ the system can be described by

$$\begin{pmatrix} \dot{x}_1(t) \\ \dot{x}_2(t) \end{pmatrix} = \begin{pmatrix} 0 & 1 \\ -\omega^2 & 0 \end{pmatrix} \begin{pmatrix} x_1(t) \\ x_2(t) \end{pmatrix} + \begin{pmatrix} 0 \\ v(t) \end{pmatrix}$$
$$y(t) = (1 \quad 0)\mathbf{x}(t) + w(t)$$

It follows from (6.21) that each element of the covariance matrix satisfies

$$\begin{aligned}
\dot{\Sigma}_{11}(t) &= 2\Sigma_{12}(t) - W^{-1}\Sigma_{11}^{2}(t) \\
\dot{\Sigma}_{12}(t) &= \Sigma_{22}(t) - \omega^{2}\Sigma_{11}(t) - W^{-1}\Sigma_{11}(t)\Sigma_{12}(t) \\
\dot{\Sigma}_{22}(t) &= -2\omega^{2}\Sigma_{12}(t) + V - W^{-1}\Sigma_{12}^{2}(t)
\end{aligned}$$

with the boundary condition $\Sigma(0) = \Sigma_0$.

The steady-state solutions for infinite t of the above equations should be non-negative definite and are given by

$$\begin{aligned}
\Sigma_{11} &= W[2(\sqrt{(\gamma)} - \omega^{2})]^{1/2} \\
\Sigma_{12} &= W[\sqrt{(\gamma)} - \omega^{2}] \\
\Sigma_{22} &= W[\gamma(\sqrt{(\gamma)} - \omega^{2})]^{1/2}
\end{aligned}$$

where $\gamma \equiv \omega^4 + V/W$.

The stationary Kalman filter is thus described by

$$\begin{aligned}
\dot{\hat{x}}_1(t) &= -[2(\sqrt{(\gamma)} - \omega^{2})]^{1/2}\hat{x}_1(t) + \hat{x}_2(t) + \{2[\sqrt{(\gamma)} - \omega^{2}]\}^{1/2}y(t) \\
\dot{\hat{x}}_2(t) &= -\sqrt{(\gamma)}\hat{x}_1(t) + [\sqrt{(\gamma)} - \omega^{2}]y(t)
\end{aligned}$$

Although the system is not asymptotically stable, the covariance matrix tends to a finite value for infinite t. The asymptotic properties of the Kalman filter are discussed in the next section.

6.2.1 Properties of the Kalman Filter

Asymptotic characteristics

We pay attention only to the stationary property of the covariance equation (6.21) because of the difficulties associated with studying directly the asymptotic behaviour of the random variable $\hat{\mathbf{x}}(t)$. The Riccati equation (6.21) which the convariance matrix $\Sigma(t)$ satisfies has a similar form to (5.5a) which was used to solve the optimal control problem presented in Chapter 5. By making comparisons of (6.21) and (5.5) it is seen that the following dual relations hold between the optimal regulator and the optimal state estimation problems:

$$\begin{aligned}
&\Sigma \leftrightarrow P, \qquad A \leftrightarrow A^{\mathrm{T}}, \qquad C^{\mathrm{T}} \leftrightarrow B \\
&W \leftrightarrow R, \qquad \Gamma V \Gamma^{\mathrm{T}} \leftrightarrow Q, \qquad K \leftrightarrow -F^{\mathrm{T}}, \\
&t \leftrightarrow t_{\mathrm{f}} - t.
\end{aligned}$$

By considering the results of the optimal regulator problem, we can obtain the asymptotic characteristics of the Kalman filter. The following theorems can be derived corresponding to Corollary 5.1 and Theorem 5.2.

Theorem 6.3

If the linear dynamical system (A, C) is observable, the solution of (6.21) for infinite t tends to a non-negative matrix Σ which satisfies

$$0 = A\Sigma + \Sigma A^{\mathrm{T}} + \Gamma V \Gamma^{\mathrm{T}} - \Sigma C^{\mathrm{T}} W^{-1} C\Sigma \tag{6.22}$$

Theorem 6.4

If the linear system (A, Γ, C) is both controllable and observable, then the solution of (6.21) tends to a unique positive definite matrix Σ and the matrix $(A - \Sigma C^{\mathrm{T}} W^{-1} C)$ is stable.

Innovations process

The innovation process $\boldsymbol{\nu}(t)$ is defined by

$$\boldsymbol{\nu}(t) = \mathbf{y}(t) - C\hat{\mathbf{x}}(t) \tag{6.23}$$

The innovations process is a forcing term for the Kalman filter of (6.19), which contributes to the operation of corrections in the state estimate. $\boldsymbol{\nu}(t)$ may also be written using (6.2) and (6.8b) as

$$\begin{aligned} \boldsymbol{\nu}(t) &= C\mathbf{x}(t) + \mathbf{w}(t) - C\hat{\mathbf{x}}(t) \\ &= C\tilde{\mathbf{x}}(t) + \mathbf{w}(t) \end{aligned} \tag{6.24}$$

We now discuss properties of the innovations process $\boldsymbol{\nu}(t)$. Since $\hat{\mathbf{x}}(t)$ is an unbiased estimate, it follows from (6.4) that

$$E\{\boldsymbol{\nu}(t)\} = 0 \tag{6.25}$$

The covariance of $\boldsymbol{\nu}(t)$ is given by

$$\begin{aligned} E\{\boldsymbol{\nu}(t)\boldsymbol{\nu}^{\mathrm{T}}(\tau)\} = {} & CE\{\tilde{\mathbf{x}}(t)\tilde{\mathbf{x}}^{\mathrm{T}}(t)\}C^{\mathrm{T}} + CE\{\tilde{\mathbf{x}}(t)\mathbf{w}^{\mathrm{T}}(\tau)\} \\ & + E\{\mathbf{w}(t)\tilde{\mathbf{x}}^{\mathrm{T}}(\tau)\}C^{\mathrm{T}} + E\{\mathbf{w}(t)\mathbf{w}^{\mathrm{T}}(\tau)\} \end{aligned} \tag{6.26}$$

It is easily shown from (6.1), (6.2) and (6.19) that the estimation error $\tilde{\mathbf{x}}(t)$ satisfies

$$\dot{\tilde{\mathbf{x}}}(t) = (A - K(t)C)\tilde{\mathbf{x}}(t) + \Gamma\mathbf{v}(t) - K(t)\mathbf{w}(t) \tag{6.27}$$

and thus $\tilde{\mathbf{x}}(t)$ does not depend on the known input $\mathbf{u}(t)$. The solution of (6.27) is

$$\tilde{\mathbf{x}}(t) = \Phi(t, t')\tilde{\mathbf{x}}(t') + \int_{t'}^{t} \Phi(t, \sigma)\{\Gamma\mathbf{v}(\sigma) - K(\sigma)\mathbf{w}(\sigma)\}\, \mathrm{d}\sigma \tag{6.28}$$

Substituting (6.28) into (6.26), and making use of the noise characteristics given from (6.3) to (6.6), we obtain

$$E\{\boldsymbol{\nu}(t)\boldsymbol{\nu}^{\mathrm{T}}(\tau)\} = C\Phi(t,\tau)\Sigma(\tau)C^{\mathrm{T}} - C\Phi(t,\tau)K(\tau)W + W\,\delta(t-\tau) \quad \text{for } t \geqslant \tau$$

Then it follows, by use of (6.20), that

$$E\{\boldsymbol{\nu}(t)\boldsymbol{\nu}^{\mathrm{T}}(\tau)\} = W\,\delta(t-\tau) \tag{6.29}$$

This implies that there is no information left in the innovations process $\boldsymbol{\nu}(t)$ if $\hat{\mathbf{x}}(t)$ is the optimal estimate.

Theorem 6.5

The innovations process defined by (6.23) is a white Gaussian process with mean and covariance matrix given by

$$E\{\boldsymbol{\nu}(t)\} = 0 \quad \text{and} \quad E\{\boldsymbol{\nu}(t)\boldsymbol{\nu}^{\mathrm{T}}(\tau)\} = W\,\delta(t-\tau) \tag{6.30}$$

6.2.2 Treatment of Various Types of Random Noise

The Kalman filter given in Theorem 6.2 assumes the conditions from (6.3) to (6.6) for the random noise signals. In this section, we make some generalizations to the Kalman filter for the case where the system disturbance is correlated with the measurement noise, and the system disturbance and measurement noise are not white. Since the known input $\mathbf{u}(t)$ plays no significant role in the Kalman filter, then we assume that $\mathbf{u}(t) \equiv 0$.

Correlated system and measurement noises

We assume here that the system disturbance $\mathbf{v}(t)$ and the measurement noise $\mathbf{w}(t)$ are correlated, so that

$$\operatorname{cov}\{\mathbf{v}(t), \mathbf{w}(\tau)\} = S\delta(t-\tau) \tag{6.31}$$

Since the derivation of the optimal filter for this case is similar to the previous derivation of Theorem 6.2 where $\mathbf{v}(t)$ and $\mathbf{w}(t)$ are uncorrelated, we present only the results.

Theorem 6.6

Assume that $\mathbf{v}(t)$ and $\mathbf{w}(t)$ are characterized by (6.3), (6.4), (6.6) and (6.31). Then the optimal state estimate $\hat{\mathbf{x}}(t)$ for the linear system described by (6.1) and (6.2) is given by

$$\dot{\hat{\mathbf{x}}}(t) = A\hat{\mathbf{x}}(t) + K(t)\{\mathbf{y}(t) - C\hat{\mathbf{x}}(t)\}, \hat{\mathbf{x}}(0) = \bar{\mathbf{x}}_0 \tag{6.32}$$

$$K(t) = (\Sigma(t)C^{\mathrm{T}} + \Gamma S)W^{-1} \tag{6.33}$$

$$\dot{\Sigma}(t) = A\Sigma(t) + \Sigma(t)A^{\mathrm{T}} + \Gamma V \Gamma^{\mathrm{T}} \\ - \{\Sigma(t)C^{\mathrm{T}} + \Gamma S\} W^{-1} \{\Sigma(t)C^{\mathrm{T}} + \Gamma S\}^{\mathrm{T}}, \Sigma(0) = \Sigma_0 \tag{6.34}$$

Example 6.3

Consider the linear system

$$\dot{x}(t) = ax(t) + v(t)$$
$$y(t) = x(t) + \rho v(t)$$

where $a < 0$ and $\rho > 0$. Let the variance of $v(t)$ be denoted by V. Then we obtain $W = \rho^2 V$ and $S = \rho V$. It follows from (6.34) that

$$\dot{\Sigma}(t) = 2\left(a - \frac{1}{\rho}\right) \Sigma(t) - \frac{1}{\rho^2 V} \Sigma^2(t)$$

Since the stationary solution for infinite t is $\Sigma = \lim_{t\to\infty} \Sigma(t) = 0$, then it will be seen that the estimate $\hat{x}(t)$ approaches the true value of $x(t)$. The stationary Kalman filter is given by

$$\dot{\hat{x}}(t) = a\hat{x}(t) + \frac{1}{\rho} [y(t) - \hat{x}(t)]$$

Coloured system disturbance

In this section we consider the optimal filter for the case when the system disturbance is not white. This disturbance can be modelled as the output of a linear dynamical system driven with a white noise input. Then, by adjoining the dynamical system associated with the disturbance generation to the original linear system, we obtain an augmented dynamical system which has a white noise input for which we can design a Kalman filter.

Instead of condition (6.3), let $\mathbf{v}(t)$ be modelled as the output of the q-dimensional linear system

$$\dot{\boldsymbol{\lambda}}(t) = M\boldsymbol{\lambda}(t) + N\boldsymbol{\xi}(t), \qquad \boldsymbol{\lambda}(0) = \boldsymbol{\lambda}_0 \tag{6.35a}$$

$$\mathbf{v}(t) = G\boldsymbol{\lambda}(t) \tag{6.35b}$$

where $\boldsymbol{\xi}(t)$ and $\boldsymbol{\lambda}_0$ are Gaussian random variables characterized by

$$E\{\boldsymbol{\lambda}_0\} = 0, \qquad E\{\boldsymbol{\lambda}_0\boldsymbol{\lambda}_0^{\mathrm{T}}\} = \Lambda_0 \tag{6.36}$$

$$E\{\boldsymbol{\xi}(t)\} = 0, \qquad E\{\boldsymbol{\xi}(t)\boldsymbol{\xi}^{\mathrm{T}}(\tau)\} = \Xi\, \delta(t-\tau) \tag{6.37}$$

$$E\{\boldsymbol{\xi}(t)\boldsymbol{\lambda}_0^{\mathrm{T}}\} = 0. \tag{6.38}$$

It is also assumed that $\boldsymbol{\xi}(t)$ is statistically independent of $\mathbf{x}(0)$ and $\mathbf{w}(t)$ and that M is a stable matrix.

For the stationary case, we discuss the covariance matrix of the noise $\mathbf{v}(t)$. Let the transition matrix of M be denoted by $\Phi(t) = e^{Mt}$, then the solution $\boldsymbol{\lambda}(t)$ of (6.35a) is given by

$$\boldsymbol{\lambda}(t) = \int_{-\infty}^{t} \Phi(t-\sigma)N\boldsymbol{\xi}(\sigma)\,\mathrm{d}\sigma = \int_{0}^{\infty} \Phi(\sigma)N\boldsymbol{\xi}(t-\sigma)\,\mathrm{d}\sigma \tag{6.39}$$

It then follows from (6.37) and (6.39) that

$$\Lambda \equiv E\{\boldsymbol{\lambda}(t)\boldsymbol{\lambda}^{\mathrm{T}}(t)\} = \int_{0}^{\infty} \Phi(\sigma)N\Xi N^{\mathrm{T}}\Phi^{\mathrm{T}}(\sigma)\,\mathrm{d}\sigma \tag{6.40}$$

Since M is a stable matrix, it satisfies the Lyapunov equation

$$\begin{aligned} M\Lambda + \Lambda M^{\mathrm{T}} &= \int_{0}^{\infty} \{Me^{M\sigma}N\Xi N^{\mathrm{T}}e^{M^{\mathrm{T}}\sigma} + e^{M\sigma}N\Xi N^{\mathrm{T}}e^{M^{\mathrm{T}}\sigma}M^{\mathrm{T}}\}\,\mathrm{d}\sigma \\ &= \int_{0}^{\infty} \frac{\mathrm{d}}{\mathrm{d}\sigma}\{e^{M\sigma}N\Xi N^{\mathrm{T}}e^{M^{\mathrm{T}}\sigma}\}\,\mathrm{d}\sigma = -N\Xi N^{\mathrm{T}} \end{aligned}$$

that is

$$M\Lambda + \Lambda M^{\mathrm{T}} + N\Xi N^{\mathrm{T}} = 0 \tag{6.41}$$

Making use of the above results, we give the correlation function matrix $\Psi_v(\tau)$ for the random noise $\mathbf{v}(t)$. It follows from (6.39) and (6.37) that

$$\begin{aligned} \Psi_v(\tau) &\equiv E\{\mathbf{v}(t+\tau)\mathbf{v}^{\mathrm{T}}(t)\} = GE\{\boldsymbol{\lambda}(t+\tau)\boldsymbol{\lambda}^{\mathrm{T}}(t)\}G^{\mathrm{T}} \\ &= G\int_{0}^{\infty}\int_{0}^{\infty} \Phi(\sigma)NE\{\boldsymbol{\xi}(t+\tau-\sigma)\boldsymbol{\xi}^{\mathrm{T}}(t-s)\}N^{\mathrm{T}}\Phi^{\mathrm{T}}(s)\,\mathrm{d}\sigma\,\mathrm{d}sG^{\mathrm{T}} \\ &= G\int_{0}^{\infty} \Phi(\tau+s)N\Xi N^{\mathrm{T}}\Phi^{\mathrm{T}}(s)\,\mathrm{d}sG^{\mathrm{T}} \\ &= G\Phi(\tau)\Lambda G^{\mathrm{T}} \qquad \text{for } \tau \geqslant 0 \end{aligned} \tag{6.42a}$$

and also

$$\Psi_v(\tau) = G\Lambda\Phi^{\mathrm{T}}(-\tau)G^{\mathrm{T}} \qquad \text{for } \tau < 0. \tag{6.42b}$$

Example 6.4 Let the random noise $v(t)$ be the output of the linear dynamical system

$$\begin{aligned} \dot{\lambda}(t) &= -a\lambda(t) + \xi(t) \\ v(t) &= g\lambda(t) \end{aligned}$$

Since $\Phi(t) = e^{-at}$ and the solution of (6.41) is $\Lambda = \Xi/2a$, then it follows from (6.42) that the correlation function of $v(t)$ is

$$\Psi_v(\tau) = (g^2\Xi/2a)e^{-a|\tau|}$$

Example 6.5 Let the random noise $v(t)$ be the output of the second-order linear system

$$\begin{pmatrix} \dot{\lambda}_1(t) \\ \dot{\lambda}_2(t) \end{pmatrix} = \begin{pmatrix} 0 & 1 \\ -\omega_n^2 & -2\zeta\omega_n \end{pmatrix} \begin{pmatrix} \lambda_1(t) \\ \lambda_2(t) \end{pmatrix} + \begin{pmatrix} 0 \\ \omega_n^2 \end{pmatrix} \xi(t)$$

$$v(t) = (1 \quad 0)\boldsymbol{\lambda}(t)$$

The transition matrix of the above linear system is

$$\Phi(t) = \begin{pmatrix} e^{-\zeta\omega_n t}\left[\cos\omega_0 t + \dfrac{\zeta\omega_n}{\omega_0}\sin\omega_0 t\right] & \dfrac{1}{\omega_0} e^{-\zeta\omega_n t}\sin\omega_0 t \\ -\dfrac{\omega_n^2}{\omega_0} e^{-\zeta\omega_n t}\sin\omega_0 t & e^{-\zeta\omega_n t}\left[\cos\omega_0 t - \dfrac{\zeta\omega_n}{\omega_0}\sin\omega_0 t\right] \end{pmatrix}$$

where $\omega_0 = \omega_n\sqrt{(1-\zeta^2)}$. The Lyapunov equation (6.41) can be written as

$$2\Lambda_{12} = 0$$

$$\Lambda_{22} - \omega_n^2\Lambda_{11} - 2\zeta\omega_n\Lambda_{12} = 0$$

$$-2\omega_n^2\Lambda_{12} - 4\zeta\omega_n\Lambda_{22} + \omega_n^4\Xi = 0$$

and the solution for Λ is $\Lambda_{11} = \omega_n\Xi/4\zeta$, $\Lambda_{12} = 0$ and $\Lambda_{22} = \omega_n^3\Xi/4\zeta$. Therefore, it follows from (6.42a) that $\Psi_v(\tau)$ is given by

$$\Psi_v(\tau) = G\Phi(\tau)\Lambda G^{\mathsf{T}} = \frac{\omega_n\Xi}{4\zeta} e^{-\zeta\omega_n\tau}\left(\cos\omega_0\tau + \frac{\zeta\omega_n}{\omega_0}\sin\omega_0\tau\right)$$

Now, by adjoining the dynamical system (6.35) associated with the noise to the original linear system (6.1) and (6.2), we can obtain the augmented linear system.

$$\begin{pmatrix} \dot{\mathbf{x}}(t) \\ \dot{\boldsymbol{\lambda}}(t) \end{pmatrix} = \begin{pmatrix} A & \Gamma G \\ 0 & M \end{pmatrix} \begin{pmatrix} \mathbf{x}(t) \\ \boldsymbol{\lambda}(t) \end{pmatrix} + \begin{pmatrix} 0 \\ N \end{pmatrix} \boldsymbol{\xi}(t) \tag{6.43}$$

$$\mathbf{y}(t) = (C \quad 0)\mathbf{x}(t) + \mathbf{w}(t) \tag{6.44}$$

We can now easily implement the Kalman filter given in Theorem 6.1 for this augmented stochastic linear system.

Coloured measurement noise

When the measurement noise $\mathbf{w}(t)$ is not white, we can similarly model the noise as the output of a stable linear dynamical system

$$\dot{\mathbf{w}}(t) = M\mathbf{w}(t) + N\boldsymbol{\xi}(t), \qquad \mathbf{w}(0) = \mathbf{w}_0 \tag{6.45}$$

where

$$E\{\mathbf{w}_0\} = 0, \qquad E\{\mathbf{w}_0\mathbf{w}_0^{\mathsf{T}}\} = W_0 \tag{6.46}$$

$$E\{\xi(t)\} = 0, \qquad E\{\xi(t)\xi^{\mathrm{T}}(\tau)\} = \Xi\delta(t-\tau) \tag{6.47}$$

$$E\{\xi(t)\mathbf{v}^{\mathrm{T}}(\tau)\} = 0, \tag{6.48}$$

$$\operatorname{cov}\{\mathbf{x}_0, \mathbf{w}(t)\} = \operatorname{cov}\{\mathbf{x}_0, \xi(t)\} = 0 \tag{6.49}$$

With these assumptions, the least squares estimate $\hat{\mathbf{x}}(t)$ is given by the following theorem.

Theorem 6.7

If the system disturbance $\mathbf{v}(t)$ and measurement noise $\mathbf{w}(t)$ are characterized by (6.3), (6.6), and (6.45) to (6.49), the optimum estimate $\hat{\mathbf{x}}(t)$ for the linear system described by (6.1) and (6.2) is given by

$$\dot{\hat{\mathbf{x}}}(t) = A\hat{\mathbf{x}}(t) + K(t)\{\dot{\mathbf{y}}(t) - M\mathbf{y}(t) - (CA - MC)\hat{\mathbf{x}}(t)\} \tag{6.50}$$

where

$$K(t) = \{\Sigma(t)(CA - MC)^{\mathrm{T}} + \Gamma V \Gamma^{\mathrm{T}} C^{\mathrm{T}}\}\Omega^{-1} \tag{6.51}$$

$$\Omega = \{C\Gamma V \Gamma^{\mathrm{T}} C^{\mathrm{T}} + N\Xi N^{\mathrm{T}}\} \tag{6.52}$$

$$\dot{\Sigma}(t) = A\Sigma(t) + \Sigma(t)A^{\mathrm{T}} + \Gamma V \Gamma^{\mathrm{T}} - K(t)\Omega K^{\mathrm{T}}(t) \tag{6.53}$$

and the initial conditions for (6.50) and (6.53) are given by

$$\hat{\mathbf{x}}(0) = \bar{\mathbf{x}}_0 + \operatorname{cov}\{\mathbf{x}_0, \mathbf{y}(0)\}\operatorname{cov}\{\mathbf{y}(0), \mathbf{y}(0)\}^{-1}\{\mathbf{y}(0) - C\bar{\mathbf{x}}_0\} \tag{6.54a}$$

$$= \bar{\mathbf{x}}_0 + \Sigma_0 C^{\mathrm{T}}\{C\Sigma_0 C^{\mathrm{T}} + W_0\}^{-1}\{\mathbf{y}(0) - C\bar{\mathbf{x}}_0\} \tag{6.54b}$$

and

$$\Sigma(0) = \Sigma_0 - \Sigma_0 C^{\mathrm{T}}\{C\Sigma_0 C^{\mathrm{T}} + W_0\}^{-1} C\Sigma_0 \tag{6.55}$$

Proof See Appendix 6(C).

Example 6.6 Consider the stochastic linear system

$$\dot{x}(t) = ax(t) + v(t)$$
$$y(t) = cx(t) + w(t)$$
$$\dot{w}(t) = mw(t) + \xi(t)$$

The measurement noise $w(t)$ that is modelled by the last equation has the autocorrelation function $\Psi_v(\tau) = (-\Xi/2m)\mathrm{e}^{m|\tau|}$. the steady-state solution of (6.53) is

$$0 = 2a\Sigma + V - c^2\{(a-m)\Sigma + V\}^2/(c^2V + \Xi)$$

which gives

$$\Sigma = \frac{1}{c^2(a-m)^2}\{a\Xi + mc^2V + \surd[(a\Xi + mc^2V)^2 + c^2(a-m)^2V\Xi]\}$$

The gain of the Kalman filter $K(t)$ is

$$K = c\{(a-m)\Sigma + V\}/(c^2 V + \Xi)$$

and the steady-state Kalman filter is described by

$$\dot{\hat{x}}(t) = a\hat{x}(t) + K\{\dot{y}(t) - my(t) - c(a-m)\hat{x}(t)\}$$

6.3 STOCHASTIC OPTIMAL CONTROL

6.3.1 Perfect State Observation

We study the stochastic problem of controlling the linear system (6.1) subject to external disturbances so as to minimize the performance index (6.7). It is assumed that all of the states can be observed perfectly, so we take no account of the measurement equation (6.2). This problem is a stochastic version of the optimal regulator problem investigated in Chapter 5.

The optimal control is given in the next theorem.

Theorem 6.8

Consider the stochastic linear system

$$\dot{\mathbf{x}}(t) = A\mathbf{x}(t) + B\mathbf{u}(t) + \Gamma\mathbf{v}(t), \qquad \mathbf{x}(0) = \mathbf{x}_0 \tag{6.1}$$

The optimal control input $\mathbf{u}^0(t)$ that minimizes the performance index

$$J = \tfrac{1}{2}E\left[\int_0^{t_f} \{\|\mathbf{x}(t)\|_Q^2 + \|\mathbf{u}(t)\|_R^2\}\,\mathrm{d}t\right] \tag{6.7}$$

is given by state feedback in the form

$$\mathbf{u}^0(t) = -R^{-1}B^{\mathrm{T}}P(t)\mathbf{x}(t) \tag{6.56}$$

The minimum value of the performance index is

$$\min\ J = \tfrac{1}{2}\|\bar{\mathbf{x}}_0\|_{P(0)}^2 + \tfrac{1}{2}\mathrm{tr}\{\Sigma_0 P(0)\} + \tfrac{1}{2}\alpha(0) \tag{6.57}$$

where $P(t)$ and $\alpha(t)$ are characterized by the differential equations

$$-\dot{P}(t) = A^{\mathrm{T}}P(t) + P(t)A + Q - P(t)BR^{-1}B^{\mathrm{T}}P(t) \tag{6.58}$$

$$-\dot{\alpha}(t) = \mathrm{tr}\{\Gamma V \Gamma^{\mathrm{T}} P(t)\} \tag{6.59}$$

with the boundary conditions $P(t_f) = 0$ and $\alpha(t_f) = \mathbf{0}$.

Proof See Appendix 6(D)

We notice that the optimal control given by (6.56) has the same form as that of the deterministic case treated in Chapter 5. However, a difference in the value of the performance index from the deterministic one is shown by the addition of the last two terms in (6.57). The second term is owing to the uncertainty of the initial state, and the third term is owing to the uncertain system disturbances.

For infinite t_f, we use the performance index

$$J = \lim_{t_f \to \infty} \frac{1}{t_f} E\left[\int_0^{t_f} \{\| \mathbf{x}(t) \|_Q^2 + \| \mathbf{u}(t) \|_R^2\} \, \mathrm{d}t\right] \tag{6.60}$$

instead of (6.7).

Example 6.7 Consider the stochastic regulator problem where the linear dynamical system and the performance index are described by

$$\dot{x}(t) = -ax(t) + bu(t) + v(t)$$

$$J = \lim_{t_f \to \infty} \frac{1}{t_f} E\left[\int_0^{t_f} \{x^2(t) + ru^2(t)\} \, \mathrm{d}t\right]$$

the stationary solution of the Riccati equation (5.58) is

$$-2aP + 1 - \frac{b^2}{r} P^2 = 0, \qquad P > 0$$

$$P = \frac{r}{b^2}\left[\sqrt{\left(a^2 + \frac{b^2}{r}\right)} - a\right]$$

and the optimal control is given by

$$u^{\circ}(t) = -\frac{1}{b}\left[\sqrt{\left(a^2 + \frac{b^2}{r}\right)} - a\right] x(t)$$

and the minimum of the performance index for infinite t_f is

$$\min J = \tfrac{1}{2} \operatorname{tr}\{\Gamma V \Gamma^{\mathrm{T}} P\} = \frac{rV}{2b^2}\left[\sqrt{\left(a^2 + \frac{b^2}{r}\right)} - a\right]$$

since as the limiting operation is taken for t_f, the term due to the uncertainty of the initial state vanishes.

Example 6.8 Consider the stochastic linear system and the performance index

$$\dot{\mathbf{x}}(t) = A\mathbf{x}(t) + \mathbf{b}u(t) + \boldsymbol{\gamma} v(t)$$

$$A = \begin{pmatrix} 0 & 1 \\ -\omega^2 & 0 \end{pmatrix}, \mathbf{b} = \begin{pmatrix} 0 \\ b_2 \end{pmatrix}, \boldsymbol{\gamma} = \begin{pmatrix} 0 \\ \gamma_2 \end{pmatrix}$$

$$J = E\left[\int_0^{t_f} \{\| \mathbf{x}(t) \|_Q^2 + ru^2(t)\} \, \mathrm{d}t\right]$$

where $Q = \operatorname{diag}(q_1, q_2)$.

It follows from (6.56) that the optimal control is

$$u(t) = -r^{-1}\mathbf{b}^{\mathrm{T}}P(t)\mathbf{x}(t)$$

$$= -\frac{b_2}{r}\{P_{12}(t)x_1(t) + P_{22}(t)x_2(t)\}$$

where each element of $P(t)$ satisfies the Riccati equation (6.58), giving

$$\dot{P}_{11}(t) - 2\omega^2 P_{12}(t) + \frac{1}{r}b_2^2 P_{12}^2(t) + q_1 = 0$$

$$\dot{P}_{12}(t) - \omega^2 P_{22}(t) + P_{11}(t) - \frac{1}{r}b_2^2 P_{12}(t)P_{22}(t) = 0$$

$$\dot{P}_{22}(t) + 2P_{12}(t) - \frac{1}{r}b_2^2 P_{22}^2(t) + q_2 = 0$$

with the boundary condition $P(t_f) = 0$.

We briefly mention here the digital calculation of the behaviour of a linear system perturbed by random noise. Consider a discrete approximation to the above linear state equation as

$$\mathbf{x}(\overline{k+1}\Delta) = (I + \Delta A)\mathbf{x}(k\Delta) + \Delta\mathbf{b}u(k\Delta) + \gamma\Delta v(k\Delta)$$

where Δ is the sampling period. It should be noted that the covariance of the discrete-time random noise $v(k\Delta)$ is related to the covariance V for $v(t)$ by

$$\operatorname{cov}\{v(k\Delta), v(l\Delta)\} = \frac{V}{\Delta}\delta_{kl}$$

Figure 6.3 shows the solution for $P(t)$ calculated numerically for the case where $\omega = 2$, $b_2 = \gamma_2 = 1$, $q_1 = 10$, $q_2 = 1$, $r = 0.01$, $t_f = 1$, $V = 1$ and $\Delta = 0.001$. The trajectories of the states $x_1(t)$ and $x_2(t)$, and the optimal control input $u(t)$ are shown in Figure 6.4. The results are also compared with the deterministic case ($v(t) \equiv 0$) shown by the smooth solid lines. It is seen that the stochastic system behaves like the corresponding deterministic system in an averaged manner.

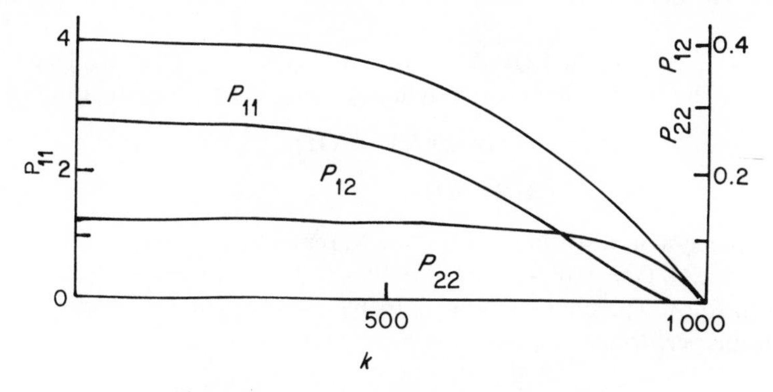

Figure 6.3 Solution for $P(t)$ of Example 6.8

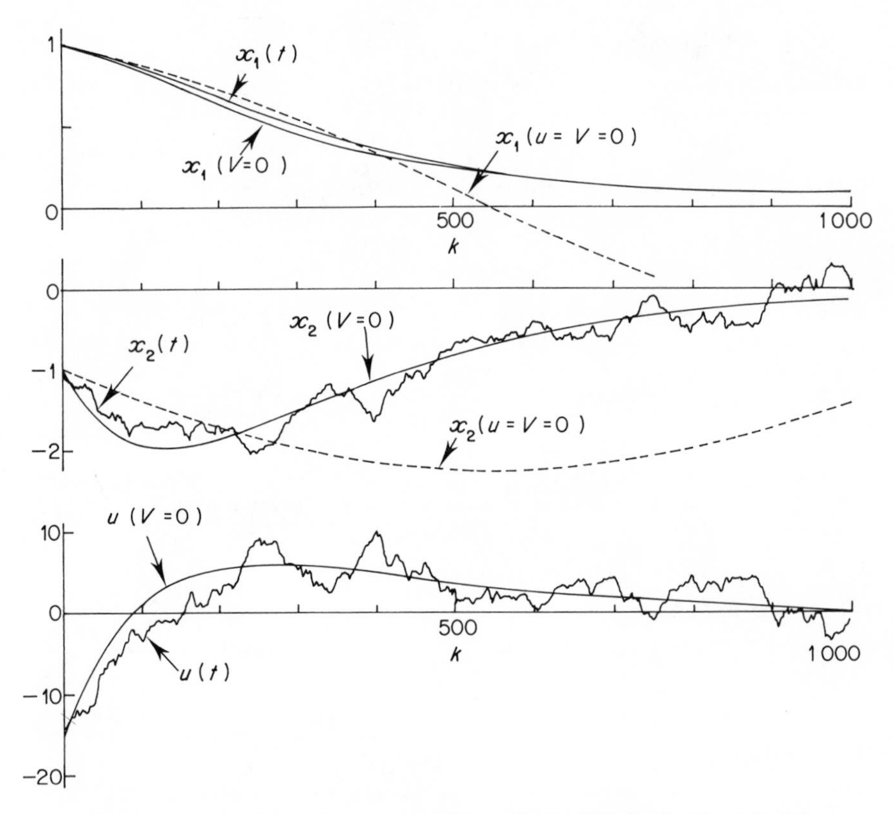

Figure 6.4 Behaviour of the state variables and control input for Example 6.8

6.3.2 The Separation Theorem

We now investigate the LQG problem stated at the beginning of the present chapter. The stochastic linear system is given by the state equation

$$\dot{\mathbf{x}}(t) = A\mathbf{x}(t) + B\mathbf{u}(t) + \Gamma\mathbf{v}(t), \qquad \mathbf{x}(0) = \mathbf{x}_0 \tag{6.1}$$

$$\mathbf{y}(t) = C\mathbf{x}(t) + \mathbf{w}(t) \tag{6.2}$$

where the system disturbance $\mathbf{v}(t)$, measurement noise $\mathbf{w}(t)$ and the initial state $\mathbf{x}_0$ are all Gaussian random variables and satisfy the assumptions from (6.3) to (6.6). The control input is to be chosen to minimize the average quadratic performance index

$$J = \tfrac{1}{2} E\left[\int_0^{t_f} \{\|\mathbf{x}(t)\|_Q^2 + \|\mathbf{u}(t)\|_R^2\}\,\mathrm{d}t\right] \tag{6.7}$$

where Q and R are non-negative-definite and positive-definite symmetric matrices respectively.

The main difference from the problem discussed in Section 6.3.1 is that information on the states can be obtained only through the output equation (6.2) and all of the states cannot be perfectly observed, so it is referred to as a partial observation problem. The problem is to choose the control input $\mathbf{u}(t)$ at time t as a function of the measured data up to time t, i.e. $\mathscr{Y}_t \equiv \{\mathbf{y}(\tau);\ 0 \leqslant \tau \leqslant t\}$ so as to minimize (6.7). Thus the controller should process the data record $\mathscr{Y}_t$ and convert it to the control input $\mathbf{u}(t)$. The separation theorem ensures that we can separate this operation into two parts: computation of the least square estimate $\hat{\mathbf{x}}(t)$ of the state $\mathbf{x}(t)$ using the input $\{\mathbf{u}(\tau); 0 \leqslant \tau \leqslant t\}$ and the output $\{\mathbf{y}(\tau); 0 \leqslant \tau \leqslant t\}$, and computation of the optimal control $\mathbf{u}(t) = F(t)\hat{\mathbf{x}}(t)$ as a linear function of the estimate $\hat{\mathbf{x}}(t)$.

Theorem 6.9 (The separation theorem)

Given the stochastic linear system (6.1) and the output equation (6.2), then the optimal control that minimizes (6.7) is given by

$$\mathbf{u}(t) = -R^{-1}B^{\mathrm{T}}P(t)\hat{\mathbf{x}}(t) \equiv F(t)\hat{\mathbf{x}}(t) \tag{6.61}$$

where $\hat{\mathbf{x}}(t)$ is the least square estimate of $\mathbf{x}(t)$ that is provided by the Kalman filter characterized by

$$\dot{\hat{\mathbf{x}}}(t) = A\hat{\mathbf{x}}(t) + B\mathbf{u}(t) + K(t)\{\mathbf{y}(t) - C\hat{\mathbf{x}}(t)\} \tag{6.19}$$

$$K(t) = \Sigma(t)C^{\mathrm{T}}W^{-1} \tag{6.20}$$

$$\dot{\Sigma}(t) = A\Sigma(t) + \Sigma(t)A^{\mathrm{T}} + \Gamma V\Gamma^{\mathrm{T}} - \Sigma(t)C^{\mathrm{T}}W^{-1}C\Sigma(t) \tag{6.21}$$

$$\text{with } \hat{\mathbf{x}}(0) = \bar{\mathbf{x}}_0 \quad \text{and} \quad \Sigma(0) = \Sigma_0$$

Then, the minimum value of the performance index is given by

$$\min\ J = \tfrac{1}{2}\,\|\,\bar{\mathbf{x}}_0\,\|^2_{P(0)} + \tfrac{1}{2}\,\mathrm{tr}\{\Sigma_0 P(0)\} + \tfrac{1}{2}\,\alpha(0) \tag{6.62}$$

where $P(t)$ and $\alpha(t)$ satisfy

$$-\dot{P}(t) = A^{\mathrm{T}}P(t) + P(t)A + Q - P(t)BR^{-1}B^{\mathrm{T}}P(t) \tag{6.63}$$

$$-\dot{\alpha}(t) = \mathrm{tr}\{Q\Sigma(t) + K(t)WK^{\mathrm{T}}(t)P(t)\} \tag{6.64}$$

with the boundary conditions $P(t_f) = 0$ and $\alpha(t_f) = 0$.

Proof It is noted that for the expectation in (6.7),

$$E\{\cdot\} = E[E\{\cdot \mid \mathscr{Y}_t\}]$$

and making use of the orthogonality of $\tilde{\mathbf{x}}(t)$ to $\mathbf{y}(\tau)$ for $0 \leqslant \tau < t$ given in (6.17) and the orthogonality of $\hat{\mathbf{x}}(t)$ to $\tilde{\mathbf{x}}(t)$, we have

$$
\begin{aligned}
&E\{\|\mathbf{x}(t)\|_Q^2 + \|\mathbf{u}(t)\|_R^2 \mid \mathscr{Y}_t\} \\
&\quad = E[\{\hat{\mathbf{x}}(t) + \tilde{\mathbf{x}}(t)\}^{\mathrm{T}} Q\{\hat{\mathbf{x}}(t) + \tilde{\mathbf{x}}(t)\} + \|\mathbf{u}(t)\|_R^2 \mid \mathscr{Y}_t] \\
&\quad = \hat{\mathbf{x}}^{\mathrm{T}}(t) Q \hat{\mathbf{x}}(t) + E\{\tilde{\mathbf{x}}^{\mathrm{T}}(t) Q \tilde{\mathbf{x}}(t)\} + \|\mathbf{u}(t)\|_R^2 \\
&\quad = \|\hat{\mathbf{x}}(t)\|_Q^2 + \|\mathbf{u}(t)\|_R^2 + \operatorname{tr}\{Q\Sigma(t)\}
\end{aligned} \tag{6.65}
$$

Since the estimation error $\tilde{\mathbf{x}}(t)$ does not depend on the control input $\mathbf{u}(t)$, the performance index (6.7) can be decomposed into the two terms

$$J = J_1 + J_2 \tag{6.66}$$

where

$$J_1 = \tfrac{1}{2} E\left[\int_0^{t_f} \{\|\hat{\mathbf{x}}(t)\|_Q^2 + \|\mathbf{u}(t)\|_R^2\}\, \mathrm{d}t\right] \tag{6.67}$$

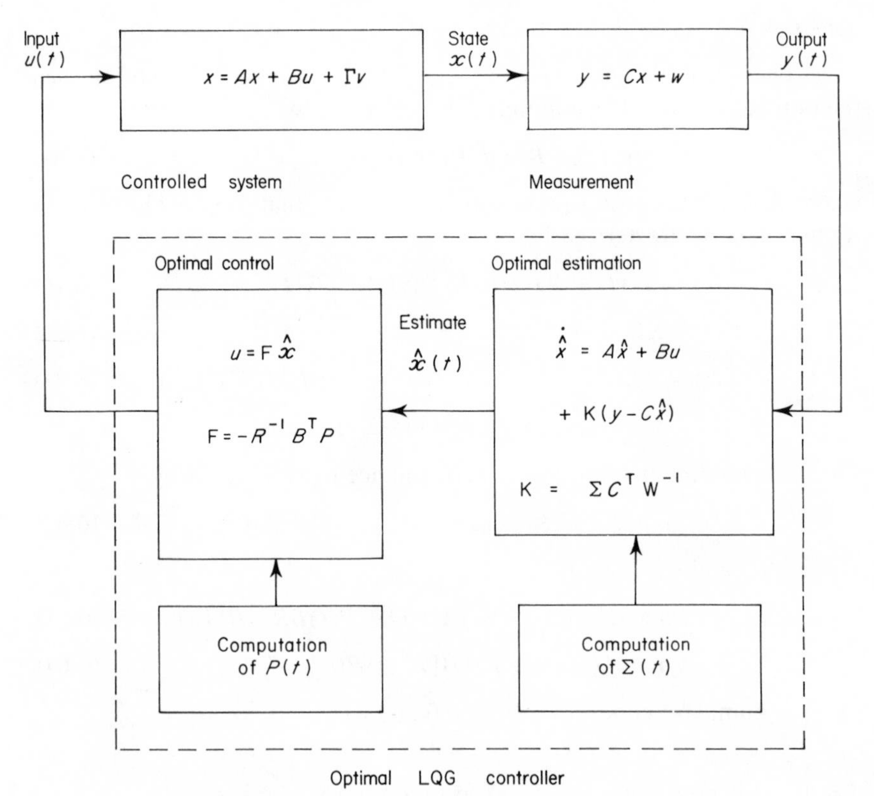

Figure 6.5 Structure of the optimal stochastic control

and

$$J_2 = \frac{1}{2}\int_0^{t_f} \mathrm{tr}\{Q\Sigma(t)\}\,\mathrm{d}t \tag{6.68}$$

The control input $\mathbf{u}(t)$ is related to J_1 only and not with J_2. Thus, the control $\mathbf{u}(t)$ should be chosen so as to minimize J_1. As mentioned previously, the estimate $\hat{\mathbf{x}}(t)$ generated by the Kalman filter can be expressed in terms of the innovations process $\boldsymbol{\nu}(t)$ as

$$\dot{\hat{\mathbf{x}}}(t) = A\hat{\mathbf{x}}(t) + B\mathbf{u}(t) + K(t)\boldsymbol{\nu}(t) \tag{6.69}$$

where the innovations process $\boldsymbol{\nu}(t)$ is white Gaussian with zero mean and covariance W. The problem of determining the optimal control so as to minimize J_1 for the system perturbed by the white noise $\boldsymbol{\nu}(t)$ is now seen to be identical to the perfect observation problem treated in Section 6.3.1, if we replace $\hat{\mathbf{x}}(t)$ with $\mathbf{x}(t)$, and $K(t)\boldsymbol{\nu}(t)$ with $\Gamma\mathbf{v}(t)$. It can be seen that the controller behaves as if $\hat{\mathbf{x}}(t)$ were the actual state $\mathbf{x}(t)$. Applying the result in Theorem 6.8, we can immediately establish Theorem 6.9.

The separation theorem states that the optimal LQG controller is separated into two parts: the optimal state estimation and the optimal stochastic controller, as illustrated in Figure 6.5. These two operations are independent in that the Kalman filter is independent of the matrices Q and R which specify the optimal controller, and the optimal control gain $F(t)$ does not depend on the statistics Γ, V, W and Σ_0 of the random noises.

Example 6.9

Assume that the stochastic linear system is the same as that treated in Example 6.8. However, here we consider the case when only the state variable $x_1(t)$ can be observed through the noise-corrupted measurement as

$$y(t) = \mathbf{c}^{\mathrm{T}}\mathbf{x}(t) + w(t)$$
$$\mathbf{c}^{\mathrm{T}} = (1 \quad 0)$$

The Kalman filter for this system has already been investigated in Example 6.2. The solution for $\Sigma(t)$ in (6.21) which was obtained numerically is shown in Figure 6.6 for the case of $W = 0.0025$. Figure 6.7 illustrates the trajectories of the actual states $x_1(t)$ and $x_2(t)$, the corresponding state estimates $\hat{x}_1(t)$ and $\hat{x}_2(t)$, and the optimal control input $u(t) = -r^{-1}\mathbf{b}^{\mathrm{T}}P(t)\hat{\mathbf{x}}(t)$. It is seen that, although the initial values of the state estimates $\hat{x}_1(0) = \hat{x}_2(0) = 0$ deviate from the true values $x_1(0) = 1$ and $x_2(0) = -1$, the estimates approach rapidly and then track their true values. Compared with Figure 6.4 where the states $x_1(t)$ and $x_2(t)$ can be observed exactly,

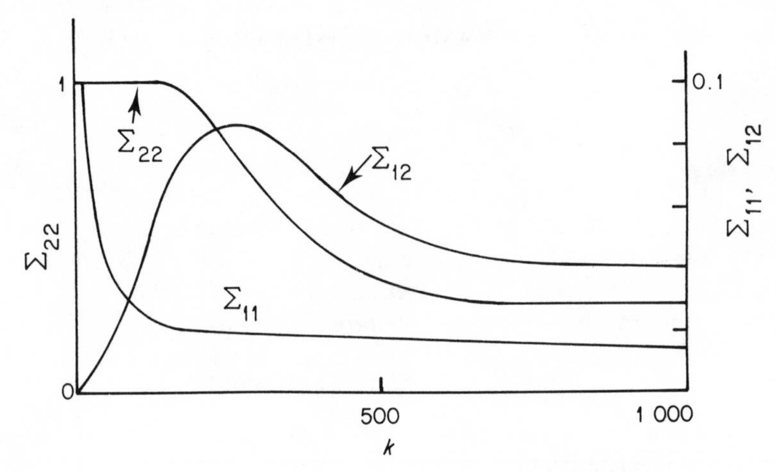

Figure 6.6 Solution for $\Sigma(t)$ for Example 6.9

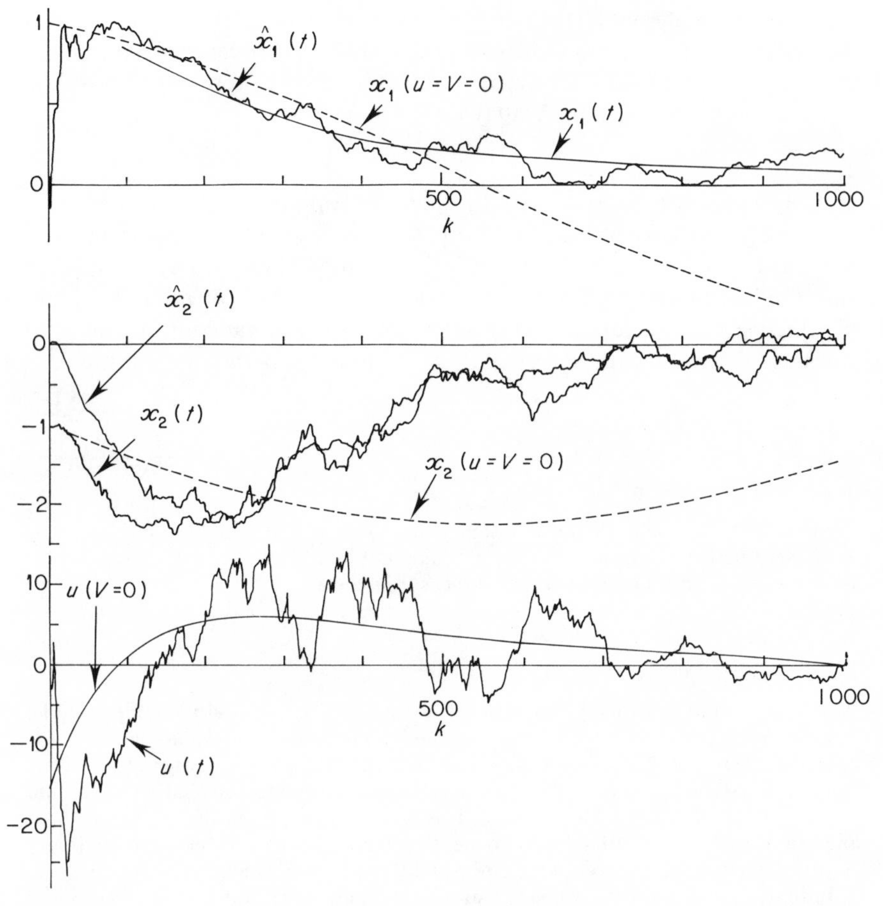

Figure 6.7 Behaviour of the state variables, state estimates and control input for Example 6.9

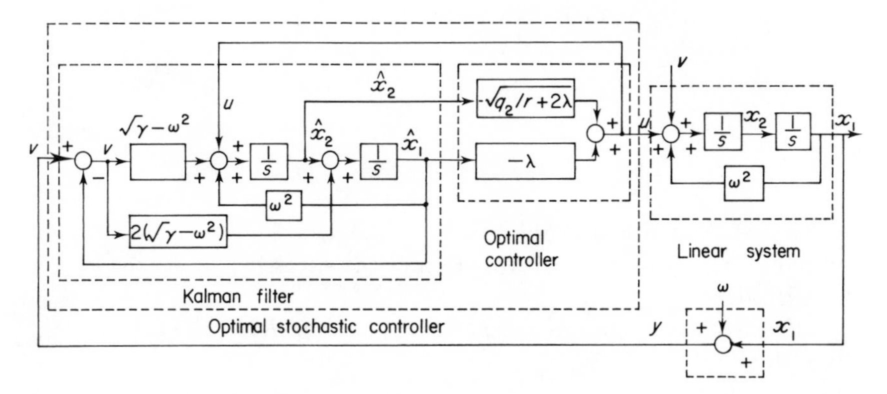

Figure 6.8 Block diagram of the stochastic optimal control system of Example 6.9

the variations in the states and control input increase slightly when only the single state $x_1(t)$ is observed through the noise-corrupted measurement.

Figure 6.8 shows the structure of the optimal LQG controller in the steady-state for $b_2 = \gamma_2 = 1$. The steady-state gain for the Kalman filter was given in Example 6.2, and the stationary optimal control is

$$u(t) = -\lambda\hat{x}_1(t) - \sqrt{(q_2/r + 2\lambda)}\hat{x}_2(t)$$

where

$$\lambda = -\omega^2 + \sqrt{(\omega^4 + q_1/r)}.$$

APPENDIX

A. Proof of Theorem 6.1

Assume the existence of the optimum kernel $H(t, \tau)$ that minimizes the mean square error I given by (6.8a). Consider another kernel $H^*(t, \tau) \equiv H(t, \tau) + \varepsilon\Xi(t, \tau)$, where the second term $\varepsilon\Xi(t, \tau)$ is an arbitrary variation from the optimum $H(t, \tau)$. Since the $H(t, \tau)$ gives the minimum of the mean square error $I\{\cdot\}$, it follows that

$$I\{H^*(t, \tau)\} \equiv I\{H(t, \tau) + \varepsilon\Xi(t, \tau)\} \geqslant I\{H(t, \tau)\}$$

Since $I\{H^*(t, \tau)\}$ can be regarded as a function of ε for a fixed $\Xi(t, \tau)$, $I\{\cdot\}$ must be stationary for $\varepsilon = 0$ and then the partial derivative of $I\{\cdot\}$ with respect to ε must vanish at $\varepsilon = 0$, i.e.,

$$\frac{\partial}{\partial\varepsilon} I\{H(t, \tau) + \varepsilon\Xi(t, \tau)\}\Big|_{\varepsilon=0} = 0 \tag{6.70}$$

Performing some manipulations for (6.70), we can obtain

$$\operatorname{tr}\left[\int_0^t \left[E\{(\mathbf{x}(t)-\bar{\mathbf{x}}(t))(\mathbf{y}(\tau)-\bar{\mathbf{y}}(\tau))^{\mathrm{T}}\} - \int_0^t H(t,\sigma) \times E\{(\mathbf{y}(\sigma)-\bar{\mathbf{y}}(\sigma))(\mathbf{y}(\tau)-\bar{\mathbf{y}}(\tau))^{\mathrm{T}}\}\,\mathrm{d}\sigma\right]\Xi(t,\tau)\,\mathrm{d}\tau\right] = 0 \quad (6.71)$$

(6.71) must hold for an arbitrary kernel matrix $\Xi(t, \tau)$. Therefore, the optimum $H(t, \tau)$ must satisfy (6.16) and the necessity of the theorem is established.

Sufficiency can be proved as follows: Evaluating $I\{H^*(t, \tau)\}$ by making use of (6.16) yields

$$I\{H^*(t,\tau)\} = I\{H(t,\tau)\} + \varepsilon^2 \operatorname{tr}\left[\int_0^t\int_0^t \Xi(t,\tau)\operatorname{cov}\{\mathbf{y}(\tau),\mathbf{y}(\sigma)\} \times \Xi^{\mathrm{T}}(t,\sigma)\,\mathrm{d}\sigma\,\mathrm{d}\tau\right]$$

It can be easily verified that the second term is positive, since W is a positive definite covariance matrix. Then we obtain $I\{H^*(t,\tau)\}\} \geqslant I\{H(t,\tau)\}$ which completes the proof.

(B) Proof of Theorem 6.2

Taking the partial derivative with respect to t in the LHS of (6.16) and making use of (6.1) and the characteristics of $\mathbf{v}(t)$ and $\mathbf{w}(t)$ given from (6.3) to (6.6), we have

$$\begin{aligned}\frac{\partial}{\partial t}\operatorname{cov}\{\mathbf{x}(t),\mathbf{y}(\tau)\} &= A\operatorname{cov}\{\mathbf{x}(t),\mathbf{y}(\tau)\} + \Gamma\operatorname{cov}\{\mathbf{v}(t),\mathbf{y}(\tau)\}\\ &= A\operatorname{cov}\{\mathbf{x}(t),\mathbf{y}(\tau)\}\end{aligned} \quad (6.72)$$

Also taking the partial derivative with respect to t on the RHS of (6.16), we have

$$\begin{aligned}\frac{\partial}{\partial t}\int_0^t & H(t,\sigma)\operatorname{cov}\{\mathbf{y}(\sigma),\mathbf{y}(\tau)\}\,\mathrm{d}\sigma\\ &= \int_0^t \frac{\partial H(t,\sigma)}{\partial t}\operatorname{cov}\{\mathbf{y}(\sigma),\mathbf{y}(\tau)\}\,\mathrm{d}\sigma + H(t,t)\operatorname{cov}\{\mathbf{y}(t),\mathbf{y}(\tau)\}\\ &= \int_0^t \frac{\partial H(t,\sigma)}{\partial t}\operatorname{cov}\{\mathbf{y}(\sigma),\mathbf{y}(\tau)\}\,\mathrm{d}\sigma + H(t,t)C\operatorname{cov}\{\mathbf{x}(t),\mathbf{y}(\tau)\}\end{aligned} \quad (6.73)$$

Substituting (6.72) and (6.73) into (6.16) gives

$$\int_0^t \left\{AH(t,\sigma) - \frac{\partial}{\partial t} H(t,\sigma) - H(t,t)CH(t,\sigma)\right\}\mathrm{cov}\{\mathbf{y}(\sigma),\mathbf{y}(\tau)\}\ \mathrm{d}\sigma = 0$$

$$\text{for } 0 \leqq \sigma < t \quad (6.74)$$

We can show by use of the positive definiteness of W that the necessary and sufficient condition for (6.74) to hold is

$$AH(t,\sigma) - \frac{\partial}{\partial t} H(t,\sigma) - K(t)CH(t,\sigma) = 0 \qquad (6.75a)$$

where

$$K(t) \equiv H(t,t) \qquad (6.75b)$$

It is obvious that (6.75a) is sufficient. The necessity can be shown as follows. Let $G(t,\sigma)$ be the LHS of (6.75b). It is seen from (6.74) and (6.16) that $H(t,\sigma) + G(t,\sigma)$ also satisfies (6.16). Then,

$$\hat{\mathbf{x}}'(t) = \hat{\mathbf{x}}(t) + \int_0^t G(t,\sigma)\{\mathbf{y}(\sigma) - \bar{\mathbf{y}}(\sigma)\}\ \mathrm{d}\sigma$$

is also the optimal estimate. Hence, the mean square erorr of $\hat{\mathbf{x}}'(t) - \hat{\mathbf{x}}(t)$ should be zero. Since

$$\int_0^t\int_0^t G(t,\sigma)\mathrm{cov}\{\mathbf{y}(\sigma),\mathbf{y}(\tau)\}G^{\mathrm{T}}(t,\tau)\ \mathrm{d}\sigma\ \mathrm{d}\tau$$

$$= \int_0^t G(t,\sigma)WG^{\mathrm{T}}(t,\sigma)\ \mathrm{d}\sigma$$

$$+ \int_0^t\int_0^t G(t,\sigma)C\ \mathrm{cov}\{\mathbf{x}(\sigma),\ \mathbf{x}(\tau)\}C^{\mathrm{T}}G^{\mathrm{T}}(t,\tau)\ \mathrm{d}\sigma\ \mathrm{d}\tau$$

and W is a positive definite covariance matrix, then we conclude that $G(t,\sigma) \equiv 0$.

Derivation of (6.19)

Differentiating (6.15) with respect to t, we have

$$\hat{\mathbf{x}}(t) = \dot{\bar{\mathbf{x}}}(t) + \int_0^t \frac{\partial}{\partial t} H(t,\tau)\{\mathbf{y}(\tau) - C\bar{\mathbf{x}}(\tau)\}\ \mathrm{d}\tau + K(t)\{\mathbf{y}(t) - C\bar{\mathbf{x}}(t)\} \qquad (6.76)$$

Since $\bar{\mathbf{x}}(t)$ satisfies

$$\dot{\bar{\mathbf{x}}}(t) = A\bar{\mathbf{x}}(t) + B\mathbf{u}(t) \qquad \text{for } \bar{\mathbf{x}}(0) = \bar{\mathbf{x}}_0, \qquad (6.77)$$

substituting (6.74) and (6.77) into (6.76) yields

$$\dot{\hat{\mathbf{x}}}(t) = A\hat{\mathbf{x}}(t) + B\mathbf{u}(t) + K(t)\{\mathbf{y}(t) - C\hat{\mathbf{x}}(t)\} \qquad \text{for } \hat{\mathbf{x}}(0) = \bar{\mathbf{x}}(0) = \bar{\mathbf{x}}_0 \tag{6.78}$$

Thus (6.19) is established.

Derivation of (6.20)

It follows from the characteristics of the random noise $\mathbf{w}(t)$ that

$$\text{cov}\{\mathbf{y}(\sigma), \mathbf{y}(\tau)\} = \text{cov}\{\mathbf{y}(\sigma), \mathbf{x}(\tau)\}C^{\mathrm{T}} + W\delta(t-\tau)$$

and

$$\text{cov}\{\mathbf{x}(t), \mathbf{y}(\tau)\} = \text{cov}\{\mathbf{x}(t), \mathbf{x}(\tau)\}C^{\mathrm{T}}$$

Then, substituting these above equations into (6.16), we obtain

$$\text{cov}\{\mathbf{x}(t), \mathbf{x}(\tau)\}C^{\mathrm{T}} = \int_0^t H(t,\sigma)\text{cov}\{\mathbf{y}(\sigma), \mathbf{x}(\tau)\}C^{\mathrm{T}}\, \mathrm{d}\sigma + H(t,\tau)W \tag{6.79}$$

and continuity of (6.79) with respect to τ gives

$$\text{cov}\{\tilde{\mathbf{x}}(t), \mathbf{x}(t)\}C^{\mathrm{T}} = K(t)W \tag{6.80}$$

By making use of the orthogonal projection lemma in (6.80), we have

$$\text{cov}\{\tilde{\mathbf{x}}(t), \mathbf{x}(t)\} = \text{cov}\{\tilde{\mathbf{x}}(t), \tilde{\mathbf{x}}(t)\} \equiv \Sigma(t)$$

Therefore since W is a positive definite matrix we obtain from (6.80)

$$K(t) = \Sigma(t)C^{\mathrm{T}}W^{-1} \tag{6.81}$$

which is the optimal gain of the Kalman filter.

Derivation of (6.21)

Subtracting (6.19) from (6.1) we obtain the equation which the estimation error $\tilde{\mathbf{x}}(t)$ should satisfy; namely

$$\dot{\tilde{\mathbf{x}}}(t) = \{A - K(t)C\}\tilde{\mathbf{x}}(t) + \Gamma\mathbf{v}(t) - K(t)\mathbf{w}(t) \tag{6.82}$$

Then the covariance matrix of the estimation error satisfies

$$\begin{aligned}
\dot{\Sigma}(t) &= E\left\{\frac{\mathrm{d}}{\mathrm{d}t}\,\tilde{\mathbf{x}}(t)\tilde{\mathbf{x}}^{\mathrm{T}}(t)\right\} + E\left\{\tilde{\mathbf{x}}(t)\,\frac{\mathrm{d}}{\mathrm{d}t}\,\tilde{\mathbf{x}}^{\mathrm{T}}(t)\right\} \\
&= \{A - K(t)C\}\,\Sigma(t) + \Gamma E\{\mathbf{v}(t)\tilde{\mathbf{x}}^{\mathrm{T}}(t)\} \\
&\quad - K(t)E\{\mathbf{w}(t)\tilde{\mathbf{x}}^{\mathrm{T}}(t)\} + \Sigma(t)\{A - K(t)C\}^{\mathrm{T}} \\
&\quad + E\{\tilde{\mathbf{x}}(t)\mathbf{v}^{\mathrm{T}}(t)\}^{\mathrm{T}} - E\{\tilde{\mathbf{x}}(t)\mathbf{w}^{\mathrm{T}}(t)\}K^{\mathrm{T}}(t)
\end{aligned} \tag{6.83}$$

Let $\Phi(t, \tau)$ be the transition matrix corresponding to $(A - K(t)C)$. Then it follows from (6.82) that

$$E\{\mathbf{v}(t)\tilde{\mathbf{x}}^{\mathrm{T}}(t)\} = E\left[\mathbf{v}(t)\left\{\Phi(t,0)\tilde{\mathbf{x}}(0) + \int_0^t \Phi(t,\tau)\{\Gamma\mathbf{v}(\tau) - K(\tau)\mathbf{w}(\tau)\}\,\mathrm{d}\tau\right\}^{\mathrm{T}}\right]$$

$$= \tfrac{1}{2} V\Gamma^{\mathrm{T}} \tag{6.84}$$

and

$$E\{\mathbf{w}(t)\tilde{\mathbf{x}}^{\mathrm{T}}(t)\} = -\tfrac{1}{2} W\Gamma^{\mathrm{T}} \tag{6.85}$$

where in the above integration we have used the formula

$$\int_0^t f(\tau)\delta(\tau - t)\,\mathrm{d}\tau = \tfrac{1}{2} f(t)$$

(C) Proof of Theorem 6.7

Let $\mathbf{z}(t)$ be the solution of the filter equation (6.50) with the initial condition (6.54). Let $\boldsymbol{\eta}(t) \equiv \mathbf{x}(t) - \mathbf{z}(t)$, then it follows from (6.1) and (6.50) that

$$\begin{aligned}\dot{\boldsymbol{\eta}}(t) &= A\mathbf{x}(t) + \Gamma\mathbf{v}(t) - A\mathbf{z}(t) - K(t)[CA\mathbf{x}(t) + C\Gamma\mathbf{v}(t) \\ &\quad + M\mathbf{w}(t) + N\boldsymbol{\xi}(t) - MC\mathbf{x}(t) - M\mathbf{w}(t) - \{CA\mathbf{z}(t) - MC\mathbf{z}(t)\}] \\ &= [A - K(t)\{CA - MC\}]\boldsymbol{\eta}(t) - K(t)N\boldsymbol{\xi}(t) + \{I - K(t)C\}\Gamma\mathbf{v}(t)\end{aligned}$$

If we define the transition matrix associated with $A - K(t)\{CA - MC\}$ by $\Phi(t, \tau)$, then we can write the solution as

$$\boldsymbol{\eta}(t) = \Phi(t,0)\boldsymbol{\eta}(0) - \int_0^t \Phi(t,\tau)K(\tau)N\boldsymbol{\xi}(\tau)\,\mathrm{d}\tau + \int_0^t \Phi(t,\tau)\{I - K(\tau)C\}\Gamma\mathbf{v}(\tau)\,\mathrm{d}\tau \tag{6.86}$$

Let $F(\tau) \equiv E\{\boldsymbol{\eta}(t)\mathbf{y}^{\mathrm{T}}(\tau)\}$ for $0 \leqslant \tau < t$. Then we have $F(0) = 0$ by use of (6.54) and the noise characteristics.

Further, it follows from (6.86) that

$$\begin{aligned}\dot{F}(\tau) &= \frac{\mathrm{d}}{\mathrm{d}\tau} E\{\boldsymbol{\eta}(t)\mathbf{x}^{\mathrm{T}}(\tau)\}C^{\mathrm{T}} + \frac{\mathrm{d}}{\mathrm{d}\tau} E\{\boldsymbol{\eta}(t)\mathbf{w}^{\mathrm{T}}(\tau)\} \\ &= E\{\boldsymbol{\eta}(t)\mathbf{x}^{\mathrm{T}}(\tau)\}A^{\mathrm{T}}C^{\mathrm{T}} + E\{\boldsymbol{\eta}(t)\mathbf{w}^{\mathrm{T}}(\tau)\}M^{\mathrm{T}} \\ &\quad + \Phi(t,\tau)[\{I - K(\tau)C\}\Gamma V\Gamma^{\mathrm{T}}C^{\mathrm{T}} - K(\tau)N\Xi N^{\mathrm{T}}]\end{aligned} \tag{6.87}$$

Since $E\{\boldsymbol{\eta}(t)\mathbf{w}^{\mathrm{T}}(\tau)\} = F(\tau) - E\{\boldsymbol{\eta}(t)\mathbf{x}^{\mathrm{T}}(\tau)\}C^{\mathrm{T}}$, then

$$\frac{\mathrm{d}}{\mathrm{d}\tau} F(\tau) = F(\tau)M^{\mathrm{T}} + \Phi(t,\tau)[E\{\boldsymbol{\eta}(\tau)\mathbf{x}^{\mathrm{T}}(\tau)\}(A^{\mathrm{T}}C^{\mathrm{T}} - C^{\mathrm{T}}M^{\mathrm{T}})$$
$$+ \Gamma V \Gamma^{\mathrm{T}} C^{\mathrm{T}} - K(\tau)\{C\Gamma V\Gamma^{\mathrm{T}}C^{\mathrm{T}} + N\Xi N^{\mathrm{T}}] \quad (6.88)$$

If $K(t)$ is chosen as (6.51) and (6.52), (6.88) becomes

$$\frac{\mathrm{d}}{\mathrm{d}\tau} F(\tau) = F(\tau)M^{\mathrm{T}} \qquad \text{for } 0 \leqslant \tau < t \quad \text{and} \quad F(0) = 0 \qquad (6.89)$$

Hence, we obtain $F(\tau) = E[\{\mathbf{x}(t) - \mathbf{z}(t)\}\mathbf{y}^{\mathrm{T}}(\tau)] = 0$ for $0 \leqslant \tau < t$, which is equivalent to the Wiener–Hopf equation. As a result, $\mathbf{z}(t)$ is the optimal estimate $\hat{\mathbf{x}}(t)$. Since $\boldsymbol{\eta}(t) = \mathbf{x}(t) - \hat{\mathbf{x}}(t) = \tilde{\mathbf{x}}(t)$, it follows from (6.86) that

$$\begin{aligned}\Sigma(t) &= E\{\tilde{\mathbf{x}}(t)\tilde{\mathbf{x}}^{\mathrm{T}}(t)\} \\ &= \Phi(t,0)E\{\tilde{\mathbf{x}}(0)\tilde{\mathbf{x}}^{\mathrm{T}}(0)\}\Phi^{\mathrm{T}}(t,0) \\ &\quad + \int_0^t \Phi(t,\tau)\, K(\tau)N\Xi N^{\mathrm{T}}K^{\mathrm{T}}(\tau)\Phi^{\mathrm{T}}(t,\tau)\,\mathrm{d}\tau \\ &\quad + \int_0^t \Phi(t,\tau)\{I - K(\tau)C\}\Gamma V\Gamma^{\mathrm{T}}\{I - K(\tau)C\}^{\mathrm{T}}\Phi^{\mathrm{T}}(t,\tau)\,\mathrm{d}\tau \end{aligned} \qquad (6.90)$$

Differentiating $\Sigma(t)$ with respect to t and making use of (6.51) and (6.52), we can conclude that (6.53) holds.

(D) Proof of Theorem 6.8

We define the optimal value of J by $\Pi(\mathbf{x}, t)$ where

$$\Pi(\mathbf{x}, t) \equiv \min_{\mathbf{u}_{[t,t_f]}} \frac{1}{2} E\left[\int_t^{t_f} \{\|\mathbf{x}(\tau)\|_Q^2 + \|\mathbf{u}(\tau)\|_R^2\}\,\mathrm{d}\tau \,|\, \mathbf{x}(t) = \mathbf{x}\right] \qquad (6.91)$$

For small Δt, $\Pi(\mathbf{x}, t)$ can be modified as

$$\begin{aligned}\Pi(\mathbf{x}, t) &= \min_{\mathbf{u}_{[t,t_f]}} \frac{1}{2} E\left[\int_t^{t+\Delta t} \{\|\mathbf{x}\|_Q^2 + \|\mathbf{u}\|_R^2\}\,\mathrm{d}\tau \right. \\ &\quad \left. + \int_{t+\Delta t}^{t_f} \{\|\mathbf{x}\|_Q^2 + \|\mathbf{u}\|_R^2\}\,\mathrm{d}\tau \,|\, \mathbf{x}(t) = \mathbf{x}\right] \\ &= \min_{\mathbf{u}_{[t,t+\Delta t]}} E\left[\frac{1}{2}\int_t^{t+\Delta t} \{\|\mathbf{x}\|_Q^2 + \|\mathbf{u}\|_R^2\}\,\mathrm{d}\tau \right. \\ &\quad \left. + \Pi(\mathbf{x}(t+\Delta t), t+\Delta t) \,|\, \mathbf{x}(t) = \mathbf{x}\right] \end{aligned} \qquad (6.92)$$

As for the system equation, we have

$$\mathbf{x}(t+\Delta t)-\mathbf{x}=\Delta\mathbf{x}=(A\mathbf{x}+B\mathbf{u})\Delta t+\Gamma\int_t^{t+\Delta t}\mathbf{v}(\tau)\,\mathrm{d}\tau+o(\Delta t)$$

Expanding $\Pi(\mathbf{x}(t+\Delta t), t+\Delta t)$ in a Taylor series gives

$$\Pi(\mathbf{x}(t+\Delta t),\ t+\Delta t)=\Pi(\mathbf{x},t)+\frac{\partial}{\partial t}\Pi(\mathbf{x},t)\ \Delta t+\Delta\mathbf{x}^{\mathrm{T}}\frac{\partial}{\partial\mathbf{x}}\Pi(\mathbf{x},t)$$

$$+\frac{1}{2}\sum_{i=1}^{n}\sum_{j=1}^{n}\frac{\partial^2}{\partial x_i\partial x_j}\Pi(\mathbf{x},t)\ \Delta x_i\ \Delta x_j+o(\Delta t)\quad(6.93)$$

Taking expectations of $\Delta\mathbf{x}$ and $\Delta\mathbf{x}\ (\Delta\mathbf{x})^{\mathrm{T}}$ yields

$$E\{\Delta\mathbf{x}\mid\mathbf{x}(t)=\mathbf{x}\}=(A\mathbf{x}+B\mathbf{u})\ \Delta t+o(\Delta t)\quad(6.94a)$$

$$E\{\Delta x_i\ \Delta x_j\mid\mathbf{x}(t)=\mathbf{x}\}=\int_t^{t+\Delta t}\int_t^{t+\Delta t}E[\{\Gamma\mathbf{v}(\tau)\}_i\{\Gamma\mathbf{v}(\sigma)\}_j]\ \mathrm{d}\tau\ \mathrm{d}\sigma+o(\Delta t)$$

$$=\{\Gamma V\Gamma^{\mathrm{T}}\}_{ij}\Delta t+o(\Delta t)\quad(6.94b)$$

Then substituting (6.93) and (6.94) into (6.92) and taking the limit as $\Delta t\to 0$, we have

$$-\frac{\partial}{\partial t}\Pi(\mathbf{x},t)=\min_{u(t)}\left[\tfrac{1}{2}\{\|\mathbf{x}\|_Q^2+\|\mathbf{u}\|_R^2\}+\{A\mathbf{x}+B\mathbf{u}\}^{\mathrm{T}}\frac{\partial}{\partial\mathbf{x}}\Pi(\mathbf{x},t)\right.$$

$$\left.+\tfrac{1}{2}\mathrm{tr}\left\{\Gamma V\Gamma^{\mathrm{T}}\frac{\partial^2}{\partial\mathbf{x}^2}\Pi(\mathbf{x},t)\right\}\right]\quad(6.95)$$

where the boundary condition is $\Pi(\mathbf{x}, t_f)=0$.

Differentiating the RHS of (6.95) with respect to $\mathbf{u}(t)$, we obtain the optimal control input

$$\mathbf{u}(t)=-R^{-1}B^{\mathrm{T}}\frac{\partial}{\partial\mathbf{x}}\Pi(\mathbf{x},t)\quad(6.96)$$

Substituting (6.96) into (6.95), we have

$$-\frac{\partial}{\partial t}\Pi(\mathbf{x},t)=\tfrac{1}{2}\mathbf{x}^{\mathrm{T}}Q\mathbf{x}+\mathbf{x}^{\mathrm{T}}A^{\mathrm{T}}\frac{\partial}{\partial\mathbf{x}}\Pi(\mathbf{x},t)+\tfrac{1}{2}\mathrm{tr}\left[\Gamma V\Gamma^{\mathrm{T}}\frac{\partial^2}{\partial\mathbf{x}^2}\Pi(\mathbf{x},t)\right]$$

$$-\tfrac{1}{2}\left\{\frac{\partial}{\partial\mathbf{x}}\Pi(\mathbf{x},t)\right\}^{\mathrm{T}}BR^{-1}B^{\mathrm{T}}\left\{\frac{\partial}{\partial\mathbf{x}}\Pi(\mathbf{x},t)\right\}\quad(6.97)$$

If the solution of (6.97) is assumed to have the form

$$\Pi(\mathbf{x},t)=\tfrac{1}{2}\mathbf{x}^{\mathrm{T}}P(t)\mathbf{x}(t)+\tfrac{1}{2}\alpha(t)\quad(6.98)$$

then we can derive the equations that $P(t)$ and $\alpha(t)$ should satisfy as

$$-\dot{P}(t) = A^{\mathrm{T}}P(t) + P(t)A + Q - P(t)BR^{-1}B^{\mathrm{T}}P(t) \tag{6.99}$$

$$-\dot{\alpha}(t) = \mathrm{tr}[\Gamma V \Gamma^{\mathrm{T}} P(t)] \tag{6.100}$$

where the boundary conditions are $P(t_f) = 0$ and $\alpha(t_f) = 0$.

It is seen by use of (6.96) and (6.98) that the optimal control is

$$\mathbf{u}(t) = -R^{-1}B^{\mathrm{T}}P(t)\mathbf{x}(t) \tag{6.101}$$

Consequently, the minimum value of J is given from (6.98) by

$$\min J = E[\tfrac{1}{2}\mathbf{x}^{\mathrm{T}}(0)P(0)\mathbf{x}(0) + \tfrac{1}{2}\alpha(0)] \tag{6.102}$$

PROBLEMS

P6.1 Find the steady state gain of the Kalman filter, when A, Γ, C, V and W are given as follows:

$$A = \begin{pmatrix} 0 & -\alpha_0 \\ 1 & -\alpha_1 \end{pmatrix}, \ \Gamma = \begin{pmatrix} 1 & 0 \\ 0 & 1 \end{pmatrix}, \ C = (0 \quad 1); \ V = \begin{pmatrix} q_1 & 0 \\ 0 & q_2 \end{pmatrix}$$

$$W = r$$

P6.2 Consider the coupled linear system

$$\begin{pmatrix} \dot{x}_1 \\ \dot{x}_2 \end{pmatrix} = \begin{pmatrix} -a_1 & \varepsilon \\ 0 & -a_2 \end{pmatrix} \begin{pmatrix} x_1 \\ x_2 \end{pmatrix} + \begin{pmatrix} v_1 \\ v_2 \end{pmatrix}$$

$$\begin{pmatrix} y_1 \\ y_2 \end{pmatrix} = \begin{pmatrix} 1 & 0 \\ 0 & 1 \end{pmatrix} \begin{pmatrix} x_1 \\ x_2 \end{pmatrix} + \begin{pmatrix} w_1 \\ w_2 \end{pmatrix}$$

where v_1, v_2, w_1 and w_2 are mutually uncorrelated white noise processes. If ε is zero, the system is composed of two independent first-order systems. Construct the Kalman filter for the coupled second-order system and compare it with that designed for the system as two independent systems.

P6.3 Derive the Kalman filter for the stochastic linear system given by

$$\dot{x}(t) = -x(t), \qquad x(0) = x_0 \text{ (uncertain)}$$

$$y(t) = x(t) + w(t)$$

$$\ddot{w}(t) = n(t)$$

where $w(t)$ is zero-mean coloured measurement noise which is the output of a double integrator with white noise $n(t)$ as input.

P6.4 Consider the stochastic linear system

$$\dot{\mathbf{x}}(t) = \begin{pmatrix} 0 & -1 \\ 1 & -2.5 \end{pmatrix} \mathbf{x}(t) + \begin{pmatrix} 1 & 0 \\ 0 & 1 \end{pmatrix} \mathbf{v}(t)$$

$$y(t) = (0 \quad 1)\mathbf{x}(t) + w(t)$$

When the steady-state gain $\mathbf{k}$ of the Kalman filter is given by

$$\mathbf{k} = \begin{pmatrix} 1 \\ 0.5 \end{pmatrix},$$

find the covariance matrix V of the system disturbance $\mathbf{v}(t)$.

P6.5 The measurement output $y(t)$ is given by

$$y(t) = c + v(t)$$

where c is an unknown constant which has a zero mean Gaussian distribution with a variance σ^2, and $v(t)$ is a white Gaussian noise with variance V. Find the Kalman filter for estimating c.

P6.6 Consider the linear stochastic system given by

$$\dot{\mathbf{x}}(t) = A\mathbf{x}(t) + B\mathbf{b} + \mathbf{v}(t)$$

$$\dot{\mathbf{b}} = 0$$

$$y(t) = C\mathbf{x}(t) + D\mathbf{b} + \mathbf{w}(t)$$

where $\mathbf{v}(t)$ and $\mathbf{w}(t)$ are white Gaussian random variables with variance V and W respectively.

(a) By adjoining $\mathbf{b}$ to $\mathbf{x}(t)$, we have the new state variable $\mathbf{z}(t)$ as

$$\mathbf{z}(t) = \begin{bmatrix} \mathbf{x}(t) \\ \mathbf{b} \end{bmatrix}$$

Find the Kalman filter giving the optimal estimate $\hat{\mathbf{z}}(t) = [(\hat{\mathbf{x}}^{\mathrm{T}}(t),\ \hat{\mathbf{b}}^{\mathrm{T}}(t)]^{\mathrm{T}}$ of $\mathbf{z}(t)$.

(b) The optimal estimate $\hat{\mathbf{x}}(t)$ can also be given by

$$\hat{\mathbf{x}}(t) = \tilde{\mathbf{x}}(t) + V_x(t)\hat{\mathbf{b}}(t)$$

where $\tilde{\mathbf{x}}(t)$ is the bias-free estimate computed as if $\mathbf{b} = 0$, and $\hat{\mathbf{b}}(t)$ is the optimal estimate of the bias $\mathbf{b}$, which are given by

$$\begin{aligned} \dot{\tilde{\mathbf{x}}}(t) &= A\tilde{\mathbf{x}}(t) + \bar{P}_x(t)CW^{-1}(y(t) - C\tilde{\mathbf{x}}(t)) \\ \dot{\hat{\mathbf{b}}}(t) &= -M(t)(V_x^{\mathrm{T}}(t)C^{\mathrm{T}} + D^{\mathrm{T}})W^{-1}(CV_x(t) + D)\hat{\mathbf{b}}(t) \\ &\quad + M(t)(V_x^{\mathrm{T}}(t)C^{\mathrm{T}}\mathrm{T} + D^{\mathrm{T}})W^{-1}(y(t) - C\tilde{\mathbf{x}}(t)) \end{aligned}$$

$$\dot{V}_x(t) = (A - \bar{P}_x(t)C^{\mathrm{T}}W^{-1}C)V_x(t) + (B - \bar{P}_x(t)C^{\mathrm{T}}W^{-1}D)$$

$$\dot{\bar{P}}_x(t) = A\bar{P}_x(t) + \bar{P}_x(t)A^{\mathrm{T}} - \bar{P}_x(t)C^{\mathrm{T}}W^{-1}C\bar{P}_x(t) + Q$$

$$\dot{M}(t) = -M(t)(V_x^{\mathrm{T}}(t)C^{\mathrm{T}} + D)W^{-1}(CV_x + D)M(t)$$

where $V_x(0) = 0$, $\bar{P}_x(0) = P_x(0)$, $M(0) = P_x(0)$. Thus, by first computing the bias-free estimate $\bar{\mathbf{x}}(t)$ and then correcting it by the quantity $V_x(t)\hat{\mathbf{b}}(t)$, we can obtain the estimate $\hat{\mathbf{x}}(t)$. Verify the above relation. (Refer to B. Friedland, Treatment of bias in recursive filtering, *IEEE Trans. Automatic Control,* Vol. AC-14, No.4, pp.359–67 (1969).)

REFERENCES

CHAPTER 1

B. D. O. Anderson and S. Vongpanitlerd, *Network Analysis and Synthesis*, Prentice-Hall (1973).

R. W. Brockett, *Finite Dimensional Linear Systems*, Addison Wesley (1969).

C. T. Chen, *Introduction to Linear System Theory*, Holt Rinehart Winston (1970).

C. A. Desoer, *Notes for a Second Course on Linear Systems*, D. Van Nostrand (1970).

F. R. Gantmacher, *The Theory of Matrices* (2 volumes), Chelsea (1959).

R. E. Kalman, P. L. Falb and M. A. Arbib, *Topics in Mathematical System Theory*, McGraw-Hill (1969).

K. Ogata, *State Space Analysis of Control Systems*, Prentice-Hall (1967).

R. A. Rohrer, *Circuit Theory: An Introduction to the State Variable Approach*, McGraw-Hill (1971).

H. H. Rosenbrock, *State Space and Multivariable Theory*, Nelson (1970).

W. A. Wolovich, *Linear Multivariable Systems*, Springer-Verlag (1974).

W. M. Wonham, *Linear Multivariable Control*, Springer-Verlag (1974).

L. Zadeh and C. Desoer, *Linear System Theory*, McGraw-Hill (1963).

CHAPTER 2

E. G. Gilbert, Controllability and observability in multivariable control systems, *SIAM J. Control*, **1**, 128–151 (1963).

R. E. Kalman, Mathematical description of linear dynamical systems, *SIAM J. Control*, **1**, 152–192 (1963).

A. Kreindler and P. E. Sarachik, On the concepts of controllability and observability of linear systems, *IEEE Trans. Automatic Control*, **AC-9**, 129–136 (1964).

L. M. Silverman and H. E. Meadows, Controllability and observability in time-variable linear systems, *SIAM J. Control*, **5**, 64–73 (1967).

L. Weiss, The concepts of differential controllability and differential observability, *J. Math. Appl.*, **10**, 442–449 (1965).

W. M. Wonham, *Linear Multivariable Control*, Springer-Verlag (1974).

CHAPTER 3

D. G. Luenberger, Canonical forms for linear multivariable systems, *IEEE* **AC-12**, 290–293 (1967).

J. Ackermann, On the synthesis of linear control systems with specified characteristics, *Automatica*, **13**, 89–94 (1977).

V. M. Popov, Invariant description of linear, time-invariant controllable systems, *SIAM J. Control*, **10**, 252–264 (1972).

J. Rissanen, Recursive identification of linear systems, *SIAM J. Control*, **9**, 420–430 (1971).

P. Brunovsky, A classification of linear controllable systems, *Kybernetika, Cislo*, 173–188 (1970).

S. H. Wang and E. J. Davison, Canonical forms of linear multivariable systems, *SIAM J. Control and Optimization*, **14**, 236–250 (1976).

D. Q. Mayne, Computational procedure for the minimal realization of transfer function matrices, *Proc. IEE*, **115**, 1363–1368 (1968).

R. E. Kalman, Irreducible realization and the degree of a rational matrix, *J. SIAM*, **13**, 520–544 (1965).

R. E. Kalman, Mathematical description of linear dynamical systems, *SIAM J. Control*, **1**, 152–192 (1963).

L. M. Silverman, Realization of linear dynamical systems, *IEEE* **AC-16**, 554–567 (1971).

W. A. Wolovich and P. L. Falb, On the structure of multivariable systems, *SIAM J. Control*, **7**, 437–451 (1969).

CHAPTER 4

J. Ackermann, On the synthesis of linear control systems with specified characteristics, *Proc. IFAC Congress* **43**, 1 (1975).

M. Aoki, On sufficient conditions for optimal stabilization policies, *Review of Economic Studies*, **XL**(1), 131–138 (1973).

P. L. Falb and W. A. Wolovich, Decoupling in the design and synthesis of multivariable control systems, *IEEE*, **AC-12**, 651–659 (1967).

K. Furuta and S. Kamiyama, State feedback and inverse system, *Int. J. Control*, **25**, 229–241 (1977).

B. Gopinath, On the control of linear multiple input–output systems, *The Bell Technical Journal*, **50**(3), 1063–1081 (1971).

A. S. Morse and W. M. Wonham, Status of noninteracting control, *IEEE* **AC-16**, 568–581 (1971).

C. R. Paul, Pole specification in decoupled systems, *Int. J. Control*, **15**, 651–664 (1972).

CHAPTER 5

B. D. O. Anderson and D. G. Luenberger, Design of multivariable feedback systems, *Proc. IEE*, **114**, 395–399 (1967).

S. Arimoto and H. Hino, Performance deterioration of optimal regulators incorporating state estimators, *Int. J. Control*, **19**, 1133–1142 (1974).

R. W. Brockett, *Finite Dimensional Linear Systems*, Addison Wesley (1969).

A. E. Bryson, Jr. and D. G. Luenberger, The synthesis of regulator logic using state variable concepts, *Proc. IEEE*, **58**, 1803–1811 (1970).
K. Furuta, S. Hara and S. Mori, A class of systems with the same observer, *IEEE*, **AC-21**, 572–576 (1976).
B. Gopinath, On the control of linear multiple input–output systems, *The Bell Technical Journal*, **50**, 1063–1081 (1971).
D. G. Luenberger, Observers for multivariable systems, *IEEE*, **AC-11**, 190–197 (1966).
H. W. Smith and E. J. Davison, Design of industrial regulators, *Proc. IEE*, **119**, 1210–1216 (1972).
R. E. Kalman, When is a linear control system optimal?, *Trans. ASME*, *Ser. D*, **86**, 51–60 (1964).
J. E. Potter, Matrix quadratic solutions, *J. SIAM*, **14**, 496–501 (1966).
W. M. Wonham, *Linear Multivariable Control*, Springer-Verlag (1974).
A. Inoue, Observers for linear multivariable systems (in Japanese), *System and Control*, **16**, 585–590 (1972).

CHAPTER 6

M. H. Davis, *Linear Estimation and Stochastic Control,* Chapman & Hall (1977).
R. S. Bucy and P. Joseph, *Filtering for Stochastic Processes with Applications to Guidance,* Wiley (1968).
B. D. O. Anderson and J. B. Moore, *Optimal Filtering,* Prentice-Hall (1979).
T. Kailath, An innovation approach to least squares estimation, Part I: linear filtering in additive white noise, *IEEE Trans. Automatic Control,* AC-13, pp. 646–55 (1968).
R. E. Kalman, A new approach to linear filtering and prediction theory, *Trans. ASME, J. Basic Eng., Ser. D,* Vol. 82, pp. 35–45 (1960).
H. J. Kushner, *Stochastic Stability and Control,* Academic Press (1967).
J. Meditch, *Stochastic Linear Estimation and Control,* McGraw-Hill (1969).
A. Sage and J. Melsa, *Estimation Theory with Applications to Communications and Control,* McGraw-Hill (1970).
R. Stratonovich, *Conditional Markov Processes and their Application to Problems of Optimal Control*, American Elsevier (1968).
W. M. Wonham, On the separation theorem of stochastic control, *SIAM J. Control,* **6**, 312–26 (1968).

INDEX